THE COSMIC OASIS

the remarkable story of **earth's biosphere**

the **cosmic** oasis

MARK WILLIAMS & JAN ZALASIEWICZ

with illustrations by Anne-Sophie Milon

OXFORD
UNIVERSITY PRESS

OXFORD
UNIVERSITY PRESS

Great Clarendon Street, Oxford, OX2 6DP,
United Kingdom

Oxford University Press is a department of the University of Oxford.
It furthers the University's objective of excellence in research, scholarship,
and education by publishing worldwide. Oxford is a registered trade mark of
Oxford University Press in the UK and in certain other countries

Impression: 1

Published in the United States of America by Oxford University Press
198 Madison Avenue, New York, NY 10016, United States of America

British Library Cataloguing in Publication Data
Data available

Library of Congress Control Number: 2022930602

ISBN 978–0–19–884587–4

DOI: 10.1093/oso/9780198845874.001.0001

Printed and bound in the UK by Clays Ltd, Elcograf S.p.A.

ACKNOWLEDGEMENTS

This book grew out of the way that our work—as palaeontologists long fascinated by all that the fossil record can tell us about past life—has evolved to focus ever more on the revolution that life on Earth's life is undergoing today, and on its uncertain prospects for the future. That a book managed to emerge out of this alarmingly wide topic was in no part due to the support, encouragement and marvellously helpful advice from Latha Menon, Jenny Nugee and their colleagues at Oxford University Press.

Among our colleagues in palaeontology and geology who have most shaped our thinking about life's history have been John Norton, Adrian Rushton, Richard Fortey, David Siveter, Dick Aldridge, Barrie Rickards, Sarah Gabbott, Tom Harvey, Tony Barnosky, Reinhold Leinfelder, Scott Wing, Alan Haywood, Alex Page, Juan Carlos Berrio, Colin Waters, Tom Wong Hearing, Thijs Vandenbroucke, Stephen Himson and Rachael Holmes. Rachael was especially kind in reading an early draft of our manuscript and suggesting many improvements. Our more recent travels beyond our discipline into far wider realms of learning about life writ large have been enabled and guided by Julia Adeney Thomas, Will Steffen, Paul Crutzen, Bruno Latour, Dipesh Chakrabarty, Jacques Grinevald, Davor Vidas, Peter Haff, Bernd Scherer, John Palmesino and Ann-Sofi Rönnskog and Marta Gasparin. It has been quite a journey of discovery, that would have been quite impossible without the kind of education that these colleagues and many others have provided.

For this book, we have worked with the artist Anne-Sophie Milon, whose contribution went far beyond providing a series of inspired images to bring a new kind of life to these pages. She helped shape and guide the text, especially when we drifted too far away from our usual shores—and indeed her marvellous drawings inspired some of our words.

And we are grateful to our families: Asih and Milana, and Kasia, Mateusz and Camille, for their support and infinite patience throughout this long journey, and to our parents, Doreen, Les, Irena and Feliks, for giving us the gift of an interest in life, writ both large and small.

Mark Williams & Jan Zalasiewicz

CONTENTS

1

EARTHRISE

When the Apollo 8 astronaut Bill Anders saw the Earth, half lit and half in darkness and more than 230,000 miles away, rise over the lunar surface, he broke away from the tight NASA schedule to snap its image. He little realized how much that image would affect its human inhabitants. Conditions were almost perfect on that day of 24 December 1968. The side of the Earth that was illuminated, in an almost unbearably poignant symphony of blue sea, brown and green land, and white ice and cloud, contrasted unforgettably with the desolation of the Moon's cratered surface. *Earthrise*, as it came to be called, symbolized, as no image before, the delicacy, fragility, and habitability of our planet as it hangs in endless and hostile space. It became a key element of human awareness of the living, and life-giving, environment that surrounds and sustains us.

Fifty years later, Anders said that he had gone to explore the Moon, but discovered the Earth. It was the space programmes that showed our planet to be an oasis of life within a largely barren cosmos, providing an understanding quite at odds with that of earlier generations of scientists. For it was not so long ago that the Moon could be reasonably supposed to support some kind of life. That celebrated astronomer Sir William Herschel used his magnificently engineered telescope to search for life on the Moon, and in 1778 thought that he had observed lunar towns. By the next century, such speculations were being shown to be increasingly improbable (though that was not to deter such scientifically literate writers as H.G. Wells from incorporating intelligent Selenite civilizations into their stories of mystery and adventure). Nevertheless, even when the NASA Moon missions were being planned, the possibility of some kind of life on the Moon—that could be threatened by arriving humans or that could be a threat to those humans—was seriously considered.

The planning committees for the Moon launch included the visionary space scientist Carl Sagan, who had written part of his PhD thesis on the possibility of organic material there—including past or present life. He thought the Moon in its early days originally had water and an atmosphere, in which many complex organic molecules were likely produced, and perhaps even life might have developed, before the weak surface gravity took its toll, and both surface water and atmosphere were lost to space. Great care should be taken with any future astronauts, he warned, so that they are not contaminated—or that they do not contaminate any delicate lunar microbial cultures that might remain.

When the astronauts arrived, they found a lifeless, almost bone-dry wasteland; debate continues on the exact meaning of 'almost', but for all biological purposes, the Moon is dry. Moreover, the Moon's origins are being increasingly seen as dramatic and fiery enough—as the condensed remains of out-flung debris when the early proto-Earth collided with the doomed Mars-sized planet Theia—to make notions of any kind of early lunar life highly improbable.

This absolute lunar sterility is now seen, paradoxically, as a potential source of clues to what kind of organic compounds might have existed in the early Solar System. After all, what better place to preserve such primitive organic compounds than a body that is absolutely devoid of life, and thereby possesses no possibility of metabolic reworking of such early organic molecules? On Earth, such molecules have long been caught up and transformed by the all-embracing web of life. And so, just beneath the Moon's surface, in places buried deeply enough by meteorite-generated rubble to provide protection from damaging cosmic and solar radiation, there may be a treasure trove of pristine, fossilized organic molecules from the dawn of our planetary system.

The space voyages showed not just that the Moon was barren, but that our nearest planetary neighbours are also probably lifeless. Mercury much resembles the Moon and is roasted on one side by the nearby Sun. Venus does not hide lush jungles beneath its thick clouds, as in the old science fiction stories, but at its surface there is a toxic and quite certainly lifeless inferno of over 400 degrees Celsius.[1] Mars, its surface long freeze-dried, is now under the continuous gaze of robot eyes on orbiting satellites and surface rovers—yet, we still do not know if relict microbe populations lie deep underground or, indeed, if they ever developed on that planet in its warmer, wetter early years. Surprisingly, some of the brightest prospects for extraterrestrial life seem to lie far beyond the asteroid belt, in salty oceans beneath the thick ice crusts of those distant moons Europa, Callisto, and Titan—yet even here, not one positive indication has yet been found: the search for extraterrestrial life is all, so far, based on hope and conjecture. The Earth's exuberant and virtually omnipresent life marks our planet as something quite special in our Solar System.

The Earth seems even more special when the planets circling other stars are taken into account. At least the planets around our Sun are arranged in well-behaved, near-circular orbits, and can neatly be demarcated into the inner (four) rocky planets and outer (five) icy and/or gaseous ones. The plethora of distant ex-oplanets now detected has thrown up surprise after surprise, such as incandescent 'hot Jupiters' with crazily looping orbits and Neptune-sized 'super Earths'. But, the number of planets that seem to be capable of supporting something approaching Earth-type life remains terribly small.[2] A wide cosmic perspective is traditionally equated with suggesting the relative insignificance of our small planet, orbiting a humdrum Sun in the outer reaches of a thoroughly average galaxy. But from what we know now, the situation seems reversed: the Earth takes on ever-greater significance as we begin to understand its astronomical context more fully. It is that rare thing in the cosmos: a living planet.

Savants of the past, when they began to consider other planets and moons, and relate them to planet Earth, had commonly thought that they would teem with life, and generations of science fiction books have invented interstellar civilizations based on such imaginings. And yet, so far, we seem to be a rare planetary gem of abundant and complex life, amid countless dead worlds of mineral and gas. The fabric of life made up of the myriad organisms with which we share the Earth, that seem so *familiar*, are anything but the local representatives of a universal constant.

We have to recalibrate our view of life, therefore, if we are to make any sense of our position in the cosmos—and also if we are to rationally approach the extraordinary changes to Earth's life that are taking place today. This task is now urgent, not least as those extraordinary changes to life's fabric have our own species as focal point, both driving the changes and being caught up within them. As we consider life itself, and how we relate to it, we might recall that humans have pondered our relationship with nature in many different cultural traditions, quite different from that of modern industrialized society, over many millennia.

There is an exploratory journey here quite as profound as that of Bill Anders, but it does not pass through the reaches of outer space. Rather, it begins deep in prehistory.

A conversation with our ancestors

It is difficult, from our vantage point now, to appreciate quite how strange and singular is our position as humans—and how strange has become the life that surrounds us. Millions of years ago, the lives of our ancestors were in many respects more normal—at least, taking as context how life as a whole has developed over these geological timescales. It is definitely 'ancestors' in the plural. There was then a rich mosaic of ape species in Europe, Africa, and Asia. The fossilized skeletons of most of these are known only from a few fragments, but some hint at changes in diet—smaller teeth, and the beginnings of an ability to walk upright—that are signposts of the talking ape to come.

But it was the australopithecines who literally followed a different path to other apes, beginning to walk fully upright across the plains of Africa. The telltale signs of this far-reaching change are preserved in fossilized footprints from a volcanic ash at Laetoli in Tanzania. The footprints—dated at 3.7 million years old—were discovered by chance in 1976, by British palaeontologist Andrew Hill, then part of Mary Leakey's archaeological team at the site. Walking home one evening, he and another colleague were playfully throwing elephant dung at each other—palaeontologists working in the field for long months need to find their amusement where they can—when Hill ducked, lost his footing, and landed headfirst onto the footprints, not just of *Australopithecus*, but of ancient antelope and rhino.[3] Extending over more than 20 metres, traces of three individuals can be seen, one apparently walking in the footsteps of another. These great apes were co-inhabiting this landscape as best they could, one small part of a stable, long-lasting, slowly evolving ecosystem among many other large mammals—and in the shadow of most of them, both literally and metaphorically.

The species that made the human-like footprints at Laetoli was possibly *Australopithecus afarensis*, an ape that roamed widely in East Africa between 4 and 3 million years ago. Many australopithecines lived in this African landscape, subsisting on fruit, nuts, root vegetables and seeds, and scavenged meat. As well as walking upright, australopithecines also made simple stone tools—mainly crude hand axes—and therefore are a distant Prometheus to human technology. One of these apes, called *Australopithecus africanus*, is thought to be the immediate ancestor of humans.

Many species of human ape have walked the landscape over the past 3 million years, living contemporaneously with other great apes. Over that time our human ancestors have left only a few fragmentary skeletons and some stone tools. They lived, one might now say, as part of nature. Then they began to change. And after a short while, geologically speaking, so did nature.

Our own species, *Homo sapiens*—at least fragments of a skeleton that palaeontologists think is our species—dates from about 300,000 years ago. Over timeframes of thousands of human generations, these skeletons had developed changes in anatomy—a larger brain case, a less sloping forehead—that suggest an increasing capacity for complex thought, together with better-worked stone tools. But we don't know if these people also whittled at pieces of wood or made patterns in the soil with a stick. That part of the human canvas remains a blank.

Homo sapiens is now the only human species, but over most of its span it shared the Earth with other kinds of human, most notably with *Homo neanderthalensis*, or the Neanderthals. There were then other human species, such as the shadowy Denisovans, originally identified by the DNA patterns from a fingerbone found in a Siberian cave. In eastern Indonesia lived the diminutive *Homo floresiensis* that newspaper editors around the world pounced upon as the embodiment of the 'hobbit' of the JRR Tolkien stories. And in the Philippines lived *Homo luzonensis*.[4] About these, we know very little, but the Neanderthal culture has yielded more evidence.

It does, indeed, appear to have been a *culture*—in contrast to early depictions of the Neanderthals as primitive and brutish.[5] Among the places they lived was the Iberian Peninsula, until as late as 37,000 years ago, protected by the barrier of the Ebro River that dissects the landscape of northern Spain. Here their culture flourished and is preserved in artefacts as diverse as painted shells and cave engravings.[6] On the southern tip of the peninsula, in Gibraltar, Neanderthal life seems to have persisted unchanged for 100,000 years, with people living in the same sea-facing caves for countless millennia, using the same flint tools, and building fires in the same hearth—and probably using that hearth over a period longer than our human civilization.

During that vast time, sitting around that hearth and feeling the heat of the fire, did Neanderthals ever speak? Did they share tales of successful hunts or of the cold and impenetrable lands to the north? The horseshoe-shaped hyoid bone in their necks is similar to ours, and it is this bone which allows us and many other animals to make sounds. Neanderthals also possess a particular gene known to be important for language, but it is still not clear if they spoke—their complex tool technology might have been learned by copying, without any need for speech. If they could have spoken, what might they have told us in the last chance for a conversation between two sentient human species? That they were fine craftspeople who used stone, but also crafted wooden spears for hunting small game and made digging sticks for rooting out tubers. That they chose wood with care, searching out spruce and boxwood.[7] That they could weave fibre to make rope.[8] That they shared with us a symbolic use of feathers—perhaps very similar to our own feather adornments. That they cooked their food and even used natural medicines like camomile.[9] They may indeed have shown us how they searched nearby forests for various foods and medicines. Mostly though, they could have shown us how they managed to live sympathetically with their environment, neither exhausting nor damaging it, for 200,000 years. The last of the Neanderthals died out 32 millennia before modern humans learned to write. Their knowledge of this world, and their possibly wholly different sentient way of seeing it, is gone forever. In that loss there is no foil for humans, no counterpoint intelligence.

Our own species either exterminated or assimilated these cohabiting humans, as the last Neanderthals died out only a few thousand years after modern humans first arrived in Europe. There is no evidence for violent conflict. Neanderthals lived in small, isolated groups, and lacking the cultural sophistication of their cousins, may simply have been unable to compete for space and food—a pattern that continues for orangutans, gorillas, and chimps today. Nevertheless, some traces of Neanderthal (and Denisovan) DNA in living humans also suggest a degree of assimilation.

With the extinction of the Neanderthals, *Homo sapiens* had achieved hegemony. The impact on life everywhere would soon be felt.

Little Maria among the oxen

From about 70,000 years ago a new set of artefacts begin to emerge within the human repertoire, left by people who were perhaps beginning to think differently about the world. This new phenomenon has been christened 'cognitive fluidity' by Reading University archaeologist Steven Mithen. It is the capacity for abstract thought and creativity that is characteristic of humanity. Without cognitive fluidity there would be no Ziggy Stardust, no Mona Lisa, no Eiffel Tower, and no Shakespeare sonnets or humorous limericks. One of the first inklings of this change was seen on the floor of the Blombos Cave in South Africa, as deliberate scratch marks on small pieces of ochre (iron-rich clay). As cognitive fluidity developed, perhaps in many different places as people interacted with their local environment,[10] it enabled humanity's first haunting depictions of nature, in caves from Europe to Indonesia.

The first discovery of ancient cave art was made by a little girl and her father, stumbling upon the Altamira cave in northern Spain. Like so many great discoveries in science and archaeology it was a matter of pure chance.[11] The cave entrance had been blocked by rocks for generations, and any folk memory of what lay within it was long forgotten. By chance, a local hunter, Modesto Cubillas Pérez, rediscovered the entrance to the cave in 1868 whilst trying to free his dog from some rocks. The cave became well known locally, and people used it to hide away from storms and as a resting place during hunting.

News of the cave's existence reached amateur archaeologist Marcelino Sanz de Sautuola who, with Cubillas Pérez, visited the cave in 1875. Marcelino, then, observed some black lines on the wall of the cave but thought nothing of them. Four years later, armed with a greater understanding of Stone Age artefacts and accompanied by his 9-year-old daughter, Maria Justina, he revisited the cave. And it was she—while her father was busy excavating for artefacts on the floor—who looked up and noticed figures of oxen, painted on the cave walls. The cave at Altamira is about a kilometre long, and its main passage is up to 6 metres high. People were present here from at least 36,000 years ago, until the entrance was sealed by a rock fall about 13,000 years ago. Many animals are drawn on the walls, most notably the bison that were Maria's 'oxen'.

The discovery was at first lampooned by academics, most notably by the leading French archaeologists Gabriel de Mortillet and Émile de Carthailhac. They would not accept that prehistoric humans could have produced such sophisticated paintings. Their influence led to the discovery being widely disparaged, even though the Sautuolas had enlisted the help of a famous Spanish palaeontologist, Juan Vilanova y Piera. Accused of forgery, or at best naivety, Marcelino Sanz de Sautuola died, a broken man, in 1888. It would be another 14 years, during which time more sites of cave art were discovered, before Carthailhac would realize his error, and publicly apologize.

What Maria and Marcelino had discovered at Altamira was part of the first flowering of humanity's depictions of the world around them. From about 40,000 years ago, whilst much of the northern hemisphere was in the fierce grip of an ice age, people as far apart as western Europe and the island of Borneo in South East Asia began to make paintings of the animals that they lived among.

Cave art in itself was not new. Some 20,000 years earlier, Neanderthal peoples of the Iberian Peninsula may have painted dots, lines, disks, and even the outlines of their hands on to cave walls,[12] though animals do not feature in that art. As cognitively modern humans began to move into western Europe, displacing the Neanderthal people, they brought with them their new imaginations represented through paintings of lions, rhinoceros, hyena, and leopards.

Why are these depictions of animals so widespread in Stone Age cultures from Europe to the Far East? This is art made by hunter-gatherer societies, who were highly dependent on these animals for food and clothing: the depictions may be an attempt to help understand the fierce, dangerous world that they were a small part of, a way of using their cognitive fluidity to imagine the minds of the animals they relied on, and thus to develop better strategies for hunting.[13] Among the figures of horses, bears, and other animals in the caves there sometimes appear images of humans that are part animal in form. These anthropomorphisms, perhaps representing shamans, suggest a blurring of the distinction between what we in the twenty-first century might naturally separate as the human and the non-human worlds.

A power over nature

The world depicted by Stone Age art as exemplified by the oxen discovered by a child's sharp eyes, seems to change about the time that the Altamira cave was sealed by a rock fall. Though people still depicted animals in cave paintings, and have continued to do so until the present, the pictures now become filled with people too, sometimes hunting animals, sometimes frolicking and dancing, and sometimes fighting and conquering other peoples.

New patterns of human settlement and construction begin to emerge, first in the eastern Mediterranean, for example at Abu Hureyra, where hunter-gatherers lived in a village of round houses about 13,500 years ago in a landscape rich in natural produce. These people were the Natufians, a sophisticated late Stone Age culture of the Levant. They hunted game, but also used wild cereals to make bread. They had also domesticated dogs.

Another glimpse of a changing relationship between people and nature can be seen at Göbekli Tepe ('potbelly hill') in Anatolia, the world's earliest site with monumental stone architecture, dating back 12,000 years. Göbekli Tepe appears to have been ceremonial rather than residential, and the many animal depictions so far discovered of lions, gazelles, foxes, spiders, and birds provide a telltale indication of how rich the local biodiversity of this landscape was, before its subjugation by humans. Investigation at the site has revealed tens of thousands of bones of many of the animals that were consumed here, especially Persian gazelle. But, so far, there is no evidence for domesticated animals or cultivation of crops. Nevertheless, people there clearly exerted a considerable influence over the non-human life around them.[14]

Göbekli Tepe lies at a crossroads, when people had developed the organizational skills to produce the monumental structures that would be the forerunners of cities. A little time after Göbekli Tepe was constructed, the first settlements we might call towns began to emerge in the fertile crescent of the near east. Some of these towns, like Jericho, remain sites of habitation to the present. Others, such as the ancient Anatolian settlement of Çatalhöyük, vanished from history 8,000 years ago. These early settlements were founded on an abundance of produce from the beginnings of animal domestication and agriculture, processes that would eventually take over the world.

Fast forward 12,000 years to the present and a recent map made by a team from the Free University in Brussels depicts how far the process of animal domestication has changed the pattern of life on Earth.[15] The map records the numbers of domesticated animals in the landscape. Over one billion cattle are shown as dense patches through South and North America. There is a swathe of cattle across the plains of North and East Africa, and similarly dense patches in India, Europe, East Asia, and Western Australia. For Brazil—the most cattle-rich country in the world—there is one cow for each of its 209 million people, in a place where there were no cows before the time of Columbus. Other domesticated animals, like sheep and goats, also form dense patches through much of Africa, Europe, Asia, and Australasia, whilst the global distribution of pigs, chickens, horses, and ducks is similarly striking. As these spread across the globe, there was less and less space for wild animals.

The domestication of animals, beginning about 12,000 years ago, seemed to change the view of humanity's place in nature. Humans began to see themselves as something separate, something in control—a control that could spill over into institutionalized brutality.

A spectacle of death

By the beginning of the current Common Era (CE), 2,000 years ago, the world's human population had grown to 300 million, with one-fifth of these living within the Roman Empire. The city of Rome itself already exceeded one million people, and not until the beginning of the nineteenth century (with the possible exception of Angkor in Cambodia, around the twelfth century) would cities—like London, Tokyo, and Beijing—reach this scale again.

Towards the end of the first century CE, the emperor Vespasian began the construction of the Colosseum in Rome, a theatre of death that would house an audience of some 50,000. For over 400 years this building was a spectacle for the slaughter of both people and animals. Perhaps the human population of a medium-sized city, like Nottingham, lost their lives here, in gruesome 'sports' designed to entertain Rome's citizens. And maybe—it is impossible to pin down an accurate figure—as many as a million animals were slaughtered. According to the Roman historian Cassius Dio, during the inauguration of the Colosseum in about 80 CE, the emperor Titus, Vespasian's son and successor, had some 9,000 animals slaughtered in the arena during a spectacle that lasted for 100 days. Later the emperor Trajan surpassed this by slaughtering another 11,000 animals over about 120 days to celebrate his victory over the Dacians in Romania. The business of procuring animals for these spectacles extended to the remotest regions of the empire, where lions and tigers would be baited with goats and lambs, or animals were frightened into stampeding herds and captured by nets. Once delivered to the arena the animals were killed by increasingly 'novel' ways to maintain the interest of the audience. This included uneven contests between an elephant or rhinoceros and a bull, and the decapitation of ostriches using crescent-shaped arrows, so that the beheaded birds would continue running around the arena.[16]

The Colosseum became a place of industrial-scale killing of many large animals, and it followed a tradition in Roman culture extending back to at least the third century BCE. The scale of this brutality resulted in wide areas of the ancient world becoming depauperate in animals. Lions disappeared from Mesopotamia, hippos from the Nile delta, and the population of aurochs was severely reduced. Large sums of money could be made from supplying exotic animals to amphitheatres, with the cost of procuring some animals running into the equivalent of millions of dollars today.[17] The industrialization of death for spectacle had become big business. Whilst some made a tidy profit from this trade, and elite Roman citizens indulged themselves in the displays, many areas of the old world suffered irreparable environmental damage.

The relationships between the spectators of the Colosseum and the wild animals dispatched for their pleasure seems very different from that visible in the animals of the cave paintings of Altamira. One is perhaps an expression of humanity trying to understand and exert some control over the animals they hunted. The other—that of the slaughter in the Colosseum—is an expression of total power over them.

Thinking through life

The kind of ritual brutality represented by the Colosseum has extended even into modern times (in fox hunting and the bull-ring, for example) as one strand within the near-pervasive use of animals and plants by humans for food, power, and pleasure. However, this kind of use, or exploitation, was not simply and exclusively utilitarian (or gratuitous). Humans have also wondered about the range of life that surrounds them and tried to understand their position within it. While some echoes of that thinking in ancient times may be glimpsed through such as the Altamira cave paintings, it is really only with the fossilization of detailed thought patterns, with the advent of the written word, that we can trace these intellectual explorations in any detail.

Much of the earliest writing, as excavated from where it was independently developed in the Middle East, China, and Mesoamerica, is strictly utilitarian, to help in the accounting of goods such as grain and oil, or to organize tax payments. But, as written language slowly developed from these simple beginnings, it could describe, and record, more complicated ideas about the lives of humans and the world they lived in.

In the western world, one major cradle of such thinking was in the cultures of ancient Greece and Rome. These were not free of violence, oppression, and abuse, of course, as we have just seen, but nevertheless, ideas developed *and were preserved* there, that went on to catalyse what we now see, for better or worse, as a modern scientific understanding of the living world. In the kind of science that we explore in this book, one individual from these times stands out as having a profound and long-lasting influence on patterns of thought.

Aristotle (384–322 BCE) did not develop his insights in obscurity. He was a pupil of that other giant of ancient Greek philosophy, Plato, in Athens, and he himself became tutor to Alexander the Great. His thinking ranged across pretty much the whole of knowledge as it then existed, from physics and astronomy, through ethics and psychology, through to economics and government, and had profound influence on both Judaeo-Islamic and Christian scholars, into Renaissance times and beyond—even though this influence was based upon only the one-third or so of his writings that survived. Among his achievements was, arguably, the invention of biology as a coherent area of knowledge.

His biology was developed upon the basis of his own detailed observations, notably the two years he spent studying the seas around Lesbos, and also incorporating information gleaned from fishermen, travellers, and others. He used these observations as a basis to reveal common patterns, and then developed explanations to explain these patterns. Thus, he recognized about 500 kinds of animals that he arranged on a scale from lower (such as insects and molluscs) to higher (such as mammals), all these being above plants. Humans, unsurprisingly, he placed at the top of the scale. This chain of life he saw not as a product of any kind of evolution (though he recognized how organisms had functional adaptions to the environment they lived in) but as a foreordained expression of a higher purpose.

This was a sophisticated body of both knowledge and practice that set the pattern for thinking about the natural world for nearly two millennia. There is an argument, too, that Aristotle was elevated to a position of unquestionable authority, such that his ideas became dogma and stifled scientific advance for much of that time—even though he himself had developed them largely via deductions from personal observations. His ideas of a hierarchical, unchanging chain of life, the product of some mysterious higher design, could be made to chime with much religious dogma, too, to amplify their authority, and to make attempts at different kinds of thinking more difficult—and more dangerous.

One break in this intellectual dam, around the time of the European Renaissance, came through the rediscovery of the work of another philosopher of the classical period, Lucretius (~99–55 BCE), who lived through turbulent and violent times of ancient Rome. Lucretius was a follower of the Greek philosophers Epicurus (341–270 BCE) and Democritus (~460–370 BCE), and devoted much of his life to reworking and developing their ideas in an extraordinary 7415-line poem *De rerum natura*, or 'On the Nature of Things'. Lucretius diverged from Aristotle in some fundamental respects, not least by saying that there was no foreordained higher purpose that determined the nature of living things, but that organisms were simply the result of natural processes and could change over time. They were assemblies of indestructible particles he called atoms (a concept from Democritus), which simply fell apart and dispersed after death, with no immortal soul to endure afterwards. And, he said, humans were not necessarily superior to animals.

Lucretius's verse, and the ideas it contains, were admired by some of his near contemporaries, such as Virgil and Horace, but his epic poem was then largely lost to view for over a millennium. Amazingly, it survived, in rare copies in monasteries (recopied by the monks from time to time, before the parchment perished) and, even more amazingly, one copy was discovered in 1417 CE by a papal scholar, Poggio Brancolini, and not destroyed by him for its dangerous and heretical views.[18] Rather it was copied once more, this time to join—and arguably to play a pivotal role in—the ferment of ideas that were to lead to the Renaissance and then the European Enlightenment of the eighteenth century, in which independent scientific thought began to flourish once more. Certainly, Lucretius and his resolutely down-to-earth and matter-based philosophy, in which random chance events played a key role, inspired some of the key figures of those times, including Macchiavelli and Goethe. His thinking would certainly have been known, too, to a man who developed a similarly down-to-earth, and yet passionately poetical, view of the natural world, as he lived a life of high adventure to discover it: Alexander von Humboldt.

Charting the anatomy of life on Earth

The great Prussian naturalist Alexander von Humboldt had, from childhood, shown an extraordinary level of interest in natural history, and as a young man he had the opportunity and resources to pursue his dream of exploring the Earth's natural wilderness. After abortive attempts to visit the West Indies, Egypt, and North Africa, Humboldt met in 1798 with the companion with whom he would reshape the Western world's understanding of the natural world. He was the French surgeon and botanist Aimé Bonpland. With the approval of King Charles IV of Spain, the two men set sail for the Spanish colonies on 5 June 1799, not returning to Europe until August 1804.

Humboldt and Bonpland arrived in mainland South America 41 days after leaving Spain, on 16 July 1799, at the city of Cumaná in modern Venezuela. Their journey through the tropical regions of the Americas was to establish a modern understanding of how interactions between air, earth, water, and life form the biosphere. The pages of their story also reveal a deep compassion for the indigenous peoples they met, and an abhorrence for the vile manner in which some Europeans treated them. Humboldt and Bonpland made careful measurements of the natural world around them. On 4 September 1799, as they began ascending the mountains of New Andalusia to visit the Chayma people, Humboldt noted the need to reduce the amount of baggage, because of the arduous terrane they were crossing. Yet they still carried 'a sextant, a dipping needle, an apparatus to determine the magnetic variation, a few thermometers and Saussure's hygrometer'.[19] As the two men travelled that day, they noticed how different rock strata controlled the style of vegetation, in that the 'vegetation was more brilliant, wherever the Alpine Limestone was covered by a quartzose sandstone', which 'contains thin strata of a blackish clay-slate', and 'these strata hinder the water from filtering into the crevices'. Humboldt and Bonpland made many such observations, noting how the altitude and rainfall controlled the style of vegetation and fauna, and how the underlying rock influenced the soil.

For the first time, Humboldt and Bonpland began to identify the connections between the distribution of animals and plants and the soil, air temperature, and the amount of rainfall. They were establishing the basis of what we now recognize as ecoregions, and they were laying the foundations for what would later become understood as the biosphere.

Passionate, energetic, lyrical yet exact in his writing, Humboldt became something of a superstar of his time, his books becoming bestsellers and also key influences on the scientists, such as Charles Darwin, who built upon his work.

Birth of the biosphere

While Humboldt and Bonpland were travelling through South America, the distinguished French savant, Jean-Baptiste Lamarck, published a book, *Hydrogéologie*. The title looks a little strange to modern eyes, as today the word 'hydrogeology' is understood as a specific subdiscipline, the study of groundwater and of water supplies. And, Lamarck himself is mainly now notorious as the man who got evolution 'wrong' by suggesting that it occurred by the transmission to offspring of changes (such as stronger muscles obtained through exercise) acquired during the parent's lifetime. Lamarck's views are typically contrasted with those of Charles Darwin and Alfred Russel Wallace, and their theory of evolution by natural selection. That is not a wholly fair, or exact, comparison, as Lamarck was a brilliant and insightful scholar, whose reflections on biological evolution substantially preceded Darwin's and Wallace's work.[20] Indeed, it was Lamarck who coined the term 'biology'.

Hydrogéologie is a wide-ranging book on how the Earth came to be as it is. It encompasses how major landmasses and ocean basins relate to each other, and how over time land could become sea, and vice versa—Lamarck inferred a very great age of the Earth, far outstripping the biblical timescale of 6,000 years. He showed that the thickness and patterns of rock strata could not have been formed by a single catastrophic deluge, but must have formed slowly and successively, by the kinds of processes—the actions of rivers, winds, and waves—that can be observed today. And, he put life at the centre of processes that have shaped the nature of the Earth's crust.

This was not just the recognition of fossils as the remains of long-dead animals and plants. Lamarck did discuss at length how dead organisms could become petrified, but went much further. He showed how life, uniquely, created all manner of complex substances—the stuff of bone, horn, shells, wood, muscle, cartilage, and other tissues—by a 'recombination' of simpler inorganic ingredients. These complex products of life then formed an integral part of the Earth's rocky carapace, in limestones, coals, mudrocks. They were further transformed underground, their structure breaking down in more 'recombinations' to approach the simpler, inorganic state of mineral matter. These mineral products could, one day, be eventually re-made into complex living tissues, for the cycle to begin again. It was a process that resembled that classical cycle of geology in which rocks shuttle endlessly between igneous, sedimentary, and metamorphic states—but in Lamarck's version, life is at its heart.

This idea of life as a fundamental force shaping not just biology, but geology and landscape, was ahead of its time—and, in truth, was ahead, then, of any kind of scientific testability. In Lamarck's day, ideas of elements and molecules were just hazily taking shape, and still rubbing shoulders with such notions as phlogiston, the proposed 'fire-element' of substances that burn. For most of the century to come, science was to develop enormously, and its specialist disciplines mushroomed, but Lamarck's part-intuitive holistic generalizations (inspired in part by Lucretius's vision) were largely forgotten. It took several decades for life as a planetary whole to get its own name, and this was first applied by Eduard Suess.

Beginning a career in palaeontology in the mid-nineteenth century, Suess made an unfortunate start. He chose to study, and publish on, some fossil graptolites (extinct forms of plankton) in Bohemia, a region then dominated by the fiercely possessive figure of Joachim Barrande, one of the grand old men of early geology. Barrande's response to this territorial intrusion was incandescent, and he immediately published a long and vituperative diatribe on the novice's work. Suess's career survived, just (luckily, he had secured a post in Vienna's museum by the time of the older man's outburst), but he switched to other fields of study, widening his interests until they encompassed mountain ranges and ultimately the geology of the whole planet. Suess became a grand old man of geology himself in the process (and in the end buried the hatchet with Barrande). In his first book, a study of the Alps in 1875, and then again in his last one, a massive five-volume study of the geological structure of the whole Earth, Suess introduced a new term: the 'biosphere'.

The new term was something of an anomaly amid all the detail of 'hard rock' geology, but it was more than a catchily-named afterthought. It complemented two other 'spheres' also coined by Suess, the lithosphere and hydrosphere, and also the atmosphere. To Suess, the biosphere represented a straightforward enough concept: a complex envelope of life growing on a planetary surface otherwise made of rock, air, and water. But when the term, and the concept, moved from Vienna to Russia in the early twentieth century, it became in the process more complex, more powerful and—largely because of the isolation of Russia after the Bolshevik Revolution—more mysterious.

One mystery is the survival of its main exponent, Vladimir Vernadsky, a scientist who stood up successively to the Tsar, Lenin, *and* Stalin to try to protect academic independence—and lived to die peacefully in his bed at the age of 82, in 1945.[21] Part of the reason for Vernadsky's survival may have been that he was an excellent and consistently useful scientist, a mineralogist and geochemist who helped catalogue and map out Russia's huge mineral resources. He spent much of his life amid teaching and organization, having a talent for both, and building large and active research teams—though that often took him into the dangerous territory of mediating between student and academic bodies that chafed at authority, and authorities of various types itching to crack down on unruly academics. Vernadsky, a believer in science and in rational government, was astute enough to recognize the realities of both, dealt with them as best as possible—and lived to tell the tale.

Amidst all of this, Vernadsky tried to pursue his personal research interests. He had a whole raft of them, that were mostly left unfinished, as he despaired of ever managing to find the time to develop and finish any kind of meaningful scientific narrative. A mineralogist and crystallographer at the outset, he also developed an interest in soils, where geological and biological processes meet, and learnt German by reading Humboldt, and English by reading Darwin, absorbing their grander-scale ideas in the process. His polyglot enthusiasms helped him travel, and he met or worked with the likes of Marie Curie, Henry Le Chatelier and Lord Rutherford, and Eduard Suess, too, in 1911. A breadth of interests is often a curse for a scientist, as developing one's own niche within a specialization is a surer way to make a reputation and a career. But somehow, for Vernadsky, it worked out. In stolen moments between his many other duties, he puzzled away at how life relates to Earth as a whole.

His ideas crystallized in the course of a set of events, in 1917, that would have been catastrophic for most people. The Tsarist government, finally exasperated by his pro-academic activities, and trying unsuccessfully to stem the tide of events that would soon lead to revolution, fired him from his university post. During the Revolution itself, Vernadsky lost his small family estate, and was forced to move to the Ukraine, in part to recover from a bout of TB. But there, while the rest of the country was in turmoil and the world was at war, he was briefly free of his normal duties. In a few weeks of intense creativity, he took Suess's concept of the biosphere and fleshed it out in 40 handwritten pages, combining the disciplines of biology and geochemistry to create a new hybrid science, of biogeochemistry: it gave a means to describe how life shapes our habitable planet.

This window of productive time was soon closed. Vernadsky was caught up in the fighting between the Red and White Russians, one of his assistants was killed, and he himself had to go into hiding. But after the fighting was over, and life settled down in a now Bolshevik Russia, Vernadsky went back to his university job, to resume his organizing and administrative duties. He found Lenin a more accommodating and rational taskmaster than the Tsar had been, and was allowed to travel; from 1922, Vernadsky started a three-year sojourn in Paris, and it was there that those 40 pages became a slim book, *La Biosphère*, published in 1925.

In this, Vernadsky paid tribute to Suess, though his interpretation went far deeper, while also harking back to, and acknowledging, the wide-ranging ideas of Lamarck. Vernadsky's biosphere was not just the sum-total of living things, but included their interactions with air, water, and rock too: life and non-life *together* were a dynamic and integral part of our planet, that modified and indeed determined the nature and composition of the other 'spheres'. Vernadsky tried to get to the essence of life, peering through the complexity and beauty of such marvels as forests and coral reefs. Life—or 'living matter', as he put it—was, he said, essentially a mechanism by which a planet collects, stores, and uses energy from the Sun, modifying itself in the process.

To Vernadsky, then, the Earth had always been a living planet in this most fundamental of ways; he did not envisage how it might have started as a lifeless globe. And, although he was aware that the specific kinds of life had changed across the geological periods and eras, to him it always fulfilled pretty much the same function and had always been present in similar amounts (albeit tiny by comparison with the bulk of rocks, air, and water that it influenced), no matter whether as microbes, dinosaurs, mammoths, algae, or palm trees. Life, he said, will quickly expand to cover a planet, so long as there is space to fill and resources to use. It cannot be held back.

Vernadsky's sweeping 'empirical generalizations' as he called them, did not always find favour with his colleagues, who were then busy cataloguing the almost never-ending detail of the Earth's chemical, physical, and biological complexities. One colleague said that he was neglecting his solid and useful researches to contemplate the 'geochemistry of the soul of the mosquito'. His ideas, furthermore, over succeeding decades, became a generally forgotten footnote in the rapidly advancing scientific mainstream, especially as Vernadsky returned to a Russia that was to become increasingly cut off, scientifically as well as politically, from the rest of the world, as the Iron Curtain descended.

Much later, his ideas were not so much revived as reinvented in the West, under a different—and considerably more provocative—name.

Gaia, the goddess of life on Earth

As a boy growing up in pre–Second World War Britain, James Lovelock had a dislike of authority that made life unhappy for him at school, while going on to university was not an option as his family was too poor. That background, he would say later, helped him in his subsequent career. Certainly, throughout his life he remained obstinately self-sufficient, and became one of those rare things, an independent scientist, not attached to any institute or university. There was obstinacy too, in his refusing, as a wartime medical researcher (following self-education via evening classes), to carry out experiments of the effect of intense heat on live rabbits—he used his own skin instead, which was painful, but seemingly effective. Clearly, he would have fitted better among the Altamira cave-painters than in the bloodthirsty crowds of the Colosseum.

In the 1960s, Lovelock was invited by NASA to work on their Mars space missions. NASA had in mind a special mission to search for life there, but Lovelock suggested that they need not waste their time and money. It would be more effective, he said, simply to examine the chemistry of the Martian atmosphere. If it was in equilibrium with the rocks of the planetary surface, then its composition would be determined by physical and chemical properties alone, and Mars would likely be lifeless. On the other hand, if life was present on Mars, it would need to abstract materials from the atmosphere, and also continually use it as a dumping ground—which should therefore result in an atmosphere that was out of equilibrium.

Together with a colleague, Lovelock used the latest analyses of the Mars atmosphere for such a test and compared these to Earth's atmosphere. Mars' thin atmosphere, mainly of carbon dioxide, was what could be expected of a small rocky planet. Earth's, on the other hand, with abundant oxygen, together with such substances as methane, was strikingly out of equilibrium, as those gases can only be maintained in such quantities by continual biological production. This experiment, aimed at another planet, focussed Lovelock's attention on the Earth.

Thinking more deeply about life's impact, Lovelock, together with the biologist Lynn Margulis, developed the idea that life not only modifies the physical and chemical conditions of a planetary surface but can do so in such a way as to maintain habitable conditions on that planet, keeping conditions 'just right' for life to endure over geological timescales. In doing so, he trod over a good deal of the same territory that Vernadsky had done almost half a century earlier, while then being completely unaware of that Russian scientist's work, or even his existence (Vernadsky's *Biosphere* was only translated into English in 1998, a labour of love in which Lynn Margulis played a prominent role).[22]

For this regulatory 'totality of life', Lovelock took a suggestion from a neighbour, the novelist William Golding, and gave it the name of Gaia, the Greek goddess who personified the Earth. The name captured the attention of many people in a way that would not be matched by a formal scientific description. The idea of Gaia also spread the more rapidly because Lovelock, ever the maverick, published the ideas not in technical scientific journals, but rather as clearly written books aimed at a general readership.

Gaia was controversial from the beginning and remains so today. It, or perhaps she, attracted much general enthusiasm, not least from people of an ecological and/or mystical persuasion, and adherents of 'New Age' ideas. But from many scientists, the concept also attracted criticism, that could shade into ridicule. By what mechanisms, they asked, could life, made up of millions of very different (and mutually competitive) species, act to stabilize and regulate the environment and climate of an entire planet? And how could it seem to act with such a goal in mind, and with such an apparent sense of *purpose*? This seemed to fly in the face of mainstream scientific common sense, and more specifically run counter to such fundamental biological theories as Darwinian evolution, which operates without any foresight.

Lovelock, who always insisted that he was a scientist, not a mystic, and that the Gaia hypothesis was a scientifically testable concept, set to work to explore plausible regulatory mechanisms. With a colleague, Andrew Watson, he developed a simple theoretical model that they called Daisyworld. Consider, they said, an Earth where the plants were all daisies—some white, and some black. Let us consider that the white daisies begin to become more abundant, and the black ones scarcer. Such a whiter world will reflect more sunlight and become colder, and the daisies will begin to die off, though black daisies will suffer less than the white ones, as these will reflect less (and absorb more) sunlight. As black daisies begin to multiply at the expense of white daisies, the world overall will darken, absorb more sunlight, and warm overall. Once it begins to get *too* warm, then the pendulum will swing the other way, black daisies beginning to suffer and white ones supplanting them, turning down planetary temperatures once more. This was a theorized biological thermostat. Could there be real-world examples?

Lovelock pointed to some obscure but common marine plankton, that released a chemical, dimethyl sulphide (DMS), into the air that acts as seeds on which minute water droplets could form: a kind of biological cloud maker. Changes in abundance of the plankton, driven among other things by climate change, would change amounts of DMS production, and hence cloud levels, and therefore affect global temperature in turn. These kinds of ideas are still being tested, but were the kind of wholly scientific cause-and-effect mechanisms that Lovelock proposed as possible components of biosphere-driven planetary regulation.

And as to the idea of Gaia being some kind of superorganism that has a 'purpose' of keeping a planet stable, Lovelock has always denied this. It is just a mechanism he insists, that happens to work for our planet, that may be compared with purely physical and chemical controls. For instance, another way of taking carbon dioxide out of the air is simply by dissolving it in rainwater to form dilute carbonic acid, and then reacting that acid with rock minerals to produce dissolved carbonate ions; these ions then accumulate in the sea, eventually being precipitated as sedimentary layers of carbonate minerals in the form of limestone rocks—an overall process known as silicate weathering.

On Earth, silicate weathering is speeded up by warm conditions (and so carbon dioxide is sequestered more quickly, limiting the warmth), and slowed down under cold climates (allowing warmth-trapping carbon dioxide to build up once more in the atmosphere). It is viewed by many scientists as *the* most important planetary thermostat, keeping down the levels of carbon dioxide in the atmosphere. Silicate weathering, Lovelock noted, is speeded up very significantly by living organisms—microbes in the soil that break down minerals for nutrients and energy. So, given that microbial populations are also climate-controlled, silicate weathering might be viewed as part of Gaia's regulatory mechanism too.

Discussions about Gaia persist to this day, though within the scientific realm they continue to be complicated by 'New Age' connotations. Lovelock's concept, though, was a central element in the rise of an area within science that is less tainted (or enhanced, to some) by such quasi-mystical elements. Earth System science treats the Earth as a unified system comprising—and being more than the sum of—its physical, chemical, and biological components.[23] Complex feedback loops involving flows of matter and energy operate between different parts—crust, oceans, atmosphere, ice masses (cryosphere), and living matter—of the Earth System, and among the *emergent properties* are such things as surface chemistry, climate state, and biodiversity levels. Most of the energy driving this complex interlocking system comes from the Sun, though a little comes from the heat of the Earth itself, while some may be obtained also from chemical reactions involving rocks.

Earth System science is deeply interdisciplinary, drawing on, among others, geology, oceanography, meteorology, biology, ecology, chemistry, glaciology, and mathematics. It is involved in analysing and describing the major global changes taking place today, and in trying to disentangle the effects of human-driven versus natural perturbations. The Earth System approach, simultaneously wide-focus, rigorous, and quantitative, allows us to consider questions such as: what might the Earth have been like, if life had never appeared on it?

Evolution of a dead Earth

Life can certainly make a difference to a planet. What would the Earth be like if it was lifeless? Would it be a case of simply removing the cloak of animals and plants from a landscape, to reveal the bare rocks beneath? A geologist should delight in such a thing, one might think. In reality, a lifeless Earth would have a very different landscape, a different set of rocks, and probably would be a most uncomfortable—if not lethal—place for a geologist to explore.

For a start, on a dead Earth, there would be no plants and no photosynthesis, and therefore no oxygen to breathe. Free oxygen (O_2) may be a gas, and one might think of it as being as light as air, and therefore seemingly weightless. In reality, the atmosphere contains over 1,000 trillion tonnes of this gas (and the oceans a further 10 billion tonnes or so more). This is hundreds of times more in mass than the weight of the thin planetary film of living matter which has, over time, produced this gas and which keeps maintaining it, even as its fierce reactivity means it is constantly used up in chemical reactions, including the ones that take place in our lungs and bloodstream as we breathe. This huge bulk of free oxygen available in the air and oceans allows us, and other animals (and plants too), to be vigorously active. And, it rusts the Earth's surface, producing a range of oxide and hydroxide minerals that did not exist in the Earth's early history. This defining component of our planet is completely dependent on the presence of life.

Life, too, helps keep atmospheric carbon dioxide levels in check. As well as amplifying the impact of silicate weathering, as microbes speed up the decay of rocks, it can sequester carbon more directly. In photosynthesis, plants absorb this gas from the air to form their tissues. Following their death, the process is reversed; during decay, carbon dioxide is released back into the atmosphere. But the reversal is not always complete. Some plant matter does not decay, but is buried deep in strata over geological timescales. So atmospheric carbon dioxide is reduced (and levels of atmospheric oxygen are maintained). As atmospheric carbon dioxide is kept in check, so also is its planetary heat-trapping effect.

A lifeless Earth, therefore, would likely have seen carbon dioxide levels build up, and surface temperatures rise. How far would this process have gone? In the Earth's early years, when the Sun shone some 30 per cent less brightly than it does today, perhaps not enough to have a catastrophic effect. But, as the Sun slowly began to crank up its thermonuclear furnace, then it is likely that a lifeless planetary surface would eventually have heated up enough to begin to evaporate the oceans. Water vapour is a powerful greenhouse gas in its own right, so the result would likely have been a runaway greenhouse effect that would have propelled the Earth into a Venus-like state, of blisteringly high surface temperatures. With oceans boiled away, the water vapour would gradually have leaked through the stratosphere and become lost to space. With no rain and no life, carbon sequestration mechanisms would stop working, and carbon dioxide would build up to yet higher levels.

On a dead, dry Earth, planetary mechanics would be different. With no water, and no slippery clays in the surface sediment, the mechanism of plate tectonics would likely grind to a halt—if it had been able to start up at all on an Earth without life. It would be replaced by some other kind of mechanics, perhaps again one akin to that of Venus, where volcanoes erupt through a 'single-plate' crust. The landscape would be utterly alien, and hostile to any kind of life that we can imagine.

It is hard to comprehend the power of the thin film of life, that can make such a difference to an entire planet. The mass of life, of all living matter on Earth, is currently about 2.5 trillion tonnes,[24] which is very roughly one ten-billionth part of the mass of the Earth's crust—which itself is thinner compared to the Earth's bulk than is the skin of an apple compared to the apple itself. And yet, this ghost-like veil of planetary life has played a key role in determining the nature of the solid substance of the Earth. Punching far, far above its weight, life on Earth likely has allowed persistence of the conditions for its own existence. By whatever means and under whatever name, Gaia (also recently termed Life, with a capital L)[25] seems to be a key ingredient of a complex, self-regulating planet.

We are part of life, or Life, one latecomer species among millions. For most of the existence of our species we lived with little impact on our environment. Then, a few tens of millennia ago *Homo sapiens,* first in Africa, and then spreading out across the old world, triggered a larger, uniquely human impact on nature, that gathered pace over time. In the past century or so, the pace of this transformation has rocketed, and we have become a species that is now quite unlike any other in the history of this planet, capable of both profoundly understanding, and profoundly altering—perhaps destroying—the web of life that surrounds us, and of which we are a part.

The complexity of life on Earth today, and the immensely long journey it has taken to arrive at this state, is the ultimate baseline by which we try to understand the revolution now taking place. It is this baseline that we explore next.

2

A SPADEFUL OF EARTH

In a corner of the drawing room of Down House, Charles Darwin's home, stands a bassoon. It might have been used to create much beautiful music, but it achieved its own tiny fragment of immortality when played for a small audience that, as Darwin described, had its own particular form of discrimination. The bassoon's low notes, he wrote, made no impression on an earthworm. Neither did the shrill notes of a whistle, nor the tones of a nearby piano 'played as loudly as possible'. But put the worm in a pot *on* the piano, and a single note struck on the piano (Darwin tried middle C, and the 'G in the line above the treble clef') will cause it instantly to retreat into its burrow. Hence, he deduced, the worm is highly sensitive to vibrations passing through the ground, but not at all to those travelling through air.

This is just one of Darwin's many-sided investigations into these lowly animals, set down in his last book, *Vegetable Mould and Earthworms*, published in 1881, just six months before his death. Lowly? Darwin made a good case that worms have a calculating intelligence, by showing that, when dragging leaves down into their burrows, they more often than not *select* the tips of the leaves to manoeuvre first into the burrow, rather than the side or stalk— by means of geometrical analysis that, being blind, they must do solely by touch. He showed the immense work that worms do under the surface by calculating how quickly objects become buried by soil—the 'vegetable leaf mould'—that these animals bring to the surface. It was a life-long interest, started more than 40 years previously, just after he had returned from his world-spanning voyage on the *Beagle*. For this, one might blame 'Uncle Jos', his future father-in-law,[1] and the son of the famous Stoke-on-Trent potter bearing the same name, Josiah Wedgwood. Uncle Jos was a pivotal influence on Darwin. It was he who persuaded Darwin's father that Charles should be allowed to join the *Beagle* expedition. That voyage successfully completed, it was Uncle Jos's observations of the slow burial of cinder-layers spread on his fields, and his inference that this burial was due to the accumulation of myriad wormcasts, that set Darwin off on this extraordinary tangent in his studies.

Earthworms were fully worthy of the great man's attention—and yet they are only one element of the extraordinarily diverse biology that lies hidden, and usually completely ignored, under our feet.[2] In the modern parlance of soil ecology, earthworms are 'megafauna', as many are more than 20 mm long, and so it is easy to observe them and (with a little ingenuity) see how they manipulate fallen leaves and react to bassoon and piano notes. In the same size category are organisms like centipedes, millipedes, and snails, and also vertebrates like moles and shrews. The gardener's more or less constant companions, one might call them. Smaller species of the same kinds of organisms persist down into the next size category where they are joined by insects such as ants: these are the macrofauna, at between 2 and 20 mm. Being 'macro', they are still big as soil animals go, visible to a sharp-eyed gardener on hands and knees.

The real menagerie turns up at the next scale down, of meso-fauna from 2 mm down to a fifth of a millimetre. Here there is a profusion of 'microarthropods'. There are springtails, for instance, not true insects but related to them, so called because they have a kind of tension-bound lever folded under their body that can be released instantly if danger threatens, to launch them into the air. Mites abound; these animals get a bad press from the kinds which are parasites, but in the soil many species act to decompose organic matter—a vital function for us all.[3] Preying on them are the ferocious-looking pseudoscorpions, and wriggling between these are tiny, translucent enchytraeid and nematode worms. Minuscule proturans—six-legged insects so inconspicuous that they were not discovered until the twentieth century—can be present in tens of thousands per square metre of soil. There are slender 'bristletail' diplurans with long antennae to match, pauropods and symphylids which both look like tiny pale centipedes, 'wheel animalcule' rotifers, and more and various denizens of this menagerie. Many are small enough, at less than a fifth of a millimetre across, to be part of the soil microfauna, where they are joined by protozoans which mainly feed on the even smaller bacteria that abound in soil, to be in turn parasitized by myriad viruses.

This extraordinarily diverse menagerie is mostly hidden from view, and stubbornly hard to investigate. Even if you are a patient and curious gardener, quite ready to neglect the dahlias and begonias as you try to understand the world they are growing from, it is an enormous challenge to come to any kind of *real* acquaintance with soil. Not only is it a Lilliputian world for us, where worms count among the megafauna, but it is opaque, three-dimensional, and simultaneously back-breakingly heavy and terribly fragile (as any gardener will know). Soil complexity seems almost infinite; its mineral content ranges from submicroscopic clay flakes to huge boulders, while its organic content is a carbon-rich stew that ranges from the living bodies of the inhabitants through the decaying remains of countless plants and animals that form the base of this subterranean food chain, to ramifying plant roots and fungi. The root-paths, the worm-engineered tunnels, and more generally the spaces between the soil particles make for a minia- turized and insanely complex metro system for other animals to travel along, one where the pathways are partly of air and partly of capillary water that wraps around the mineral grains: while springtails and mites travel along the air passages, just next to them are animals such as microscopic rotifers that swim through the water films around the grains.

In one sense it is a sheltered world, being insulated from the extremes of heat, cold, wet, and dry that mark the surface world, just a few tens of centimetres above, but it is now known to be not in the least a steady or uniform one. It is intensely hetero- geneous geometrically, not just between the different soil layers from surface to bedrock but also laterally, as the microenviron- ments among a tangle of tree roots are quite different from the adjacent ones that the roots do not reach. It also changes through time, like the weather above. Soil ecologists have developed con- cepts such as temporary 'hot spots' and 'hot moments' of activity (that might encompass just a gram of soil), perhaps as a packet of nutrients is released from an insect carcass. Soil metabolism is often measured via its microbial activity. In what has been called a 'sleeping beauty paradox', soil processes act more slowly, below their potential—until they are awoken by a 'Prince Charm- ing' in the form of, say, the movement of a passing worm or the percolation of water from a rain shower, to bring the microbial communities into contact with their nutrient supplies.[4]

It's no wonder that Darwin, able to glimpse just a fraction of this teeming variety, compared the soil, in the last paragraph of his last book, with the coral reefs that were the subject of his very first book. The reefs are also icons of Earth's extraordinary biodiversity—albeit ones which are ostentatiously *visible* in the clear tropical waters, through the simple device of a snorkel and facemask (devices which, alas, were not available to Darwin).

We may wonder at the origin of this fabulous diversity in nature. But how do we even measure its teeming abundance?

Cataloguing the diversity of life

There are probably more than 5 million species of insects[5] on Earth and, within that vast animal group, 1.5 million species of beetles alone. There are 100,000 species of tiny diatoms—a uni-cellular alga that makes a skeleton of silica, and an important primary producer in the oceans—and 34,000 species of fish, consuming those diatoms, other fish, and invertebrates. At the opposite end of species diversity there are only about 500 species of primates—all of the monkeys, apes, lemurs, and tarsiers—and even fewer documented species of velvet worms, perhaps 200.[6] Cataloguing all of this diversity, living and fossil, according to their biological relationships, is the mission of the taxonomist.

Already by the fourth century BCE, Aristotle had distinguished animals from plants, and different groups of animals according to whether they possessed blood—such as birds, snakes, and fish—or apparently lacked blood, like snails, worms, and crabs. Today, using classification systems that assemble organisms according to their evolutionary relationships, taxonomy has managed to formally describe about 1.5 million species of life, a feat achieved mostly over the past three centuries. Even then, by some esti-mates this is no more than 15 per cent of the species on Earth.

The basic unit of taxonomy is the species, the members of which can interbreed, and share a singular set of charac-teristics determined by their genes. A taxonomist calls these characteristics a 'diagnosis'. A species of crab, for example, might be diagnosed by the shape of its legs and the types of ornamental bumps on its external skeleton, and these features would differentiate it from all other crab species. Although taxonomy sounds quite simple, in practice it needs very detailed knowledge of a whole group of organisms to identify the precise distinguishing features between one species and another. The subtle physical differences between a Bornean and Sumatran Orangutan, for example, might only be obvious to a trained primatologist.

Above the level of species in taxonomic classification is 'genus'. Each genus will have a diagnosis too, and it will comprise species that are closely related, and very similar in anatomy. The genus *Felis*, for example, contains your domestic cat, the sand cat of the North African and Middle Eastern deserts, the Chinese Mountain cat, and several other species. Above genus, the relationships become more distant and are separated by more time for evolution—often representing many millions of years. The Family Felidae contains your domestic cat, and also tigers, lions, and jaguars. Your pet, though, has been travelling a different path of evolution from the big cats for some 11 million years.

It follows that the taxonomic distinction of a cat from a dog is quite easy, because the two are quite distantly related. They are separate mammal families and occupy different parts of the mammalian evolutionary tree; their histories have evolved separately for 50 million years. Nevertheless, they are part of the same order (the Carnivora) of the mammalian class, which in turn is part of the phylum Chordata (roughly approximating to animals with backbones). For that reason, the distinction of a frog from a dandelion is even easier, because they are separated by hundreds of millions of years of evolution—going back to the Precambrian—and they belong in two separate 'Kingdoms' in taxonomic parlance, one being an animal and the other a plant.

This kind of cataloguing has proved remarkably effective in navigating life's extraordinary abundance. This is a good start, for when we try to work out where it all came from.

Early days

In the beginning, so the story goes, there was one species, the first living organism to emerge from a soup of chemical reactions on a planetary surface, and it gave rise to all that is alive today. Darwin worked this out,[7] from realizing that organisms had common features, and then daring to suggest that, over time, they had evolved and diversified into the myriad and marvellously various life-forms that we see today. We now know that the ultimate and fundamental common feature they share—that must have been possessed too by our Last Universal Common Ancestor (known as LUCA to those who ponder it closely)—is DNA, deoxyribonucleic acid. It is a chemical compound so complex and singular that it would be entirely reasonable to suggest that it appeared, and became core to a living being's fundamental machinery, only once.

This origin of all species took place somewhere between 3.5 and 4 billion years ago (or perhaps even a little earlier). It probably took place in something like the 'little warm pond' of Darwin's suggestion, or maybe by a volcanic vent on the sea floor. Somewhere, in any case, containing the raw materials of life: carbon, hydrogen, nitrogen, phosphorus, water, and so on, enough energy to make them react together (but not so much as to fry them, or blow them apart), and plenty of time (many millions of years), for *exactly* the right combination to arrive by molecular trial and error. This probably took place on Earth, but we can't completely rule out the possibility that we, our pet cats and dogs, and the flowers and sparrows in our gardens, are all aliens by origin.

In the first billion years of the existence of the Solar System, it seems likely that Mars, and maybe also Venus, were mild and habitable planets with both land and some form of ocean at the surface—before Mars froze, and Venus, victim of runaway greenhouse warming, boiled away its oceans. In those billion years, life might have arisen on one or other of our neighbours (or even on more distant planetary bodies of our Solar System, to be blasted off the surface by asteroid impact and in turn arrive on Earth on an incoming meteorite. This is perhaps unlikely—but not impossible. If life is eventually found on Mars, or on one of Jupiter's or Saturn's moons, it will be intriguing to see if it is based on DNA, or on some other molecule that serves the same purpose of faithful replication during reproduction.

But whether LUCA was home-grown or a visitor from afar, once it appeared on the Earth's surface, it did what living things do—deployed some sort of chemical energy source to power its activity, and produced copies of itself, so that when its inevitable demise came by age or accident, its replicas would continue, metabolizing and reproducing in turn. So, if LUCA is happily existing on the planet that it has made home, how did other forms arise? Why should there be a diversity of life?

For a number of reasons. The Earth is a large and varied planet. As today, the early Earth would probably have had land and sea. And even in the oceans, likely the main habitat for the earliest life, there would have been areas of deep and shallow sea, of warmer and colder water, of fast currents and gentle ones, of different kinds of rocky and sediment-covered floor that can provide nutrients. All of these different environments would have particular forms of life best suited to them, and each would cope better in its own niche, than a single form that tries to accommodate itself to all conditions. And in any setting, there are different kinds of raw materials and different energy sources, which would suit some life-forms but not others.

In DNA lies the source of the variety that might take advantage of these different settings and different conditions. When it replicates, it normally does so perfectly. But, now and again, there are errors in the reproduction, which mostly give rise to a kind of biological gobbledygook that produces an organism less fit (or totally unfit) to survive. But occasionally, one of those errors hits the jackpot, endowing the organism with some different or improved trait that gives it a better chance of survival.

What might LUCA have looked like? One might envisage it as a form of microbe, a microscopic blob of cytoplasm with strands of DNA scattered within it, and an outer membrane to protect it from the world outside. The cytoplasm also contains ribosomes, which are in effect tiny protein-making factories. Here, the proteins are built of amino acids, carefully assembled in the right order by strands of 'messenger' ribonucleic acid (RNA) which transmit the constructional blueprint from the DNA to the ribosome. And that's it—no nucleus or other internal organelles as in our own cells. Just enough to build itself, to metabolize and, when it has grown large enough, to reproduce by the simple procedure of splitting into two.

What came before LUCA? Scientists talk about an RNA world[8] that might have been a forerunner of the living world governed by DNA, a kind of transitional world of non-living but complex organic chemistry in which some of the self-organizing, metabolizing, and reproductive properties of life as we know it might have spontaneously appeared and disappeared, probably many times before the whole package of life came together, and persisted, in LUCA. RNA is a simpler chemical than is DNA, being made of just a single molecular strand. But, because of this, RNA is much more fragile, and less effective at preserving and transmitting information faithfully. Once the iconic 'double helix' of DNA arose, with its strong ladder-like structure of paired spiral strands, it became the unchallenged foundation of life on Earth, although various forms of RNA carried on running vital errands within the cells.

So, what then?

The variety of microbes

For Darwin, the long stretch of time of the early Earth, that is still informally called the Precambrian, was a mystery, one compounded by the seemingly sudden appearance of obvious, large, complex fossils in the rocks of the Cambrian Period. We now know that this 'Cambrian explosion', which took place a little over half a billion years ago, was not quite so sudden, being stretched out over tens of millions of years. And, we know that there are indeed fossils in Precambrian rock strata that extend some 3 billion years further back still, to within a billion—and perhaps half a billion—years of the Earth's origin.

Most of those Precambrian fossils are microbial. The trouble is, they tell us very little of the nature of those microbes. The signature fossils of the Precambrian are stromatolites, accumulations of fine, curved rock layers that can grow to be the size of a cannon ball. These are made by sticky microbial biofilms that trap sediment, but they can be hard to tell apart from sediment layers which are produced simply by physical means, with no involvement from living organisms.[9] Even when they are demonstrably microbial, they tell us something about the microbes' stickiness vis-à-vis sediment, but very little about their biology otherwise.

Most stromatolites—and indeed most rock strata of any kind—do not preserve the microscopic structures of Precambrian microbes. But even when a few special rock types do preserve this kind of microscopic detail (the ultra-fine-grained, silica-based rocks known as cherts usually provide the best hunting grounds for such minute fossils), then what is seen are usually tiny round or tubular blobs, some joined together in chains. The generally accepted record of these kinds of micro-fossils extends back to about 3.5 billion years ago, similar to the record for widely accepted microbial stromatolites. All of that tells us very little about what was inside those cell walls: these ancient microfossils generally look like bacteria today, but little more can be worked out from them.

Mind you, looking at modern bacteria under the microscope, even a powerful one, is not so revealing either. Until a few decades ago, classical techniques of microscopy and bacterial culturing had laboriously gathered evidence for a few thousand bacterial species, thought to be pretty much all that represented the 'prokaryotic' domain of life—that is, those single-celled organisms that lack nuclei. This contrasted with the obvious and dazzling diversity shown by the other recognized domain, of eukaryotes, these being organisms with nucleus-bearing cells that include all plants, animals, fungi, and protozoans. Small wonder that the enormous time span of the Precambrian was thought to be simple and dull biologically, with its prokaryotic microbial populations in which evolution crept along at glacial pace.

Then came a revolution that was to turn this picture on its head. It came not from a new treasure trove of fossils, or even from closer microscopic examination of living bacteria. The study here was at a much finer level, that of the molecular signatures of life. The breakthrough was spearheaded by Carl Woese, an American physicist. His interest in biology had been scant as a student, but he was later persuaded to move towards biophysics. The encouragement was inspired, though Woese always worked with a physicist's reflexes, rather than those of a biologist. That might explain why, when he published his breakthrough paper in 1977, not many biologists noticed—and of those that did, most scorned the results.[10]

The 1977 paper is short—just three pages—cogent, elegantly written, and persuasive. Perhaps it is the single illustration that helped fuel the opposition. It is simply a list of 13 names, including a yeast, a common pondweed, and several bacteria, including the type that causes diphtheria, and a few with the odd ability to convert carbon dioxide to methane. Against each is a row of 12 numbers—and that's all: no graphs, no technical drawings or interpretative diagrams. The numbers are cursorily explained, but give no hint of the backbreaking and soul-destroying work that went into gaining them. Each is a measure of how similar or different is the chemical structure of one part of one strand of RNA (one gene, effectively) within ribosomes from each of the 13 organisms studied.

Today, genetic signatures can be obtained at the flick of a switch, but at the very beginning of the science of such fingerprinting, the technology was crude (including highly radioactive phosphorus, hundred-litre tanks of kerosene, and yards of photographic film—a safety officer's nightmare). The results—smudgy dots on the thousands of sheets of film—appeared quite unintelligible to anyone who was not as determined, and as analytically minded, as Woese. He draped every space of his office with the smudgy film sheets and stared at the patterns 'all day and every day', as one of his colleagues recalled. It took him years of this obsessive study until he worked out how to extract meaningful information from them. Small wonder that he acquired a reputation as a crank, something else that did not help acceptance of his ideas.

Synthesized into this single table, the results showed that the RNA structure of the yeast and pondweed was quite dissimilar to that of the typical bacteria, including the diphtheria-causing one. That made sense and fitted the ideas of the day. But what was not expected was that the methane-generating microbes showed a pattern of numbers that was not only different from the eukaryotes but was just as different from that of the typical bacteria. To Woese, that showed that these methane-generating microbes were therefore just as distantly related to bacteria as they, and bacteria, were to eukaryotes. Therefore, he said, life on Earth was not made up of two fundamental domains: it was made of three, the third being of what he called 'archaebacteria', later shortened to archaea.

It should have been a thunderbolt, quite on the scale of the discovery of DNA's double helix. But, as was pointedly observed afterwards, most biologists did not notice. Of those that did, most—and especially the eminent ones—thought the idea nonsense, and said so loudly and brutally. How could these unfamiliar, chemically derived numbers be of any more use than the classical methods at classifying microbes? The classical approach had admittedly been a failure, and the question of microbial relationships and evolution was regarded as inherently intractable. How could this new kind of information be any better? The great and good of biology simply did not believe Woese's assertion that the RNA patterns were effective guides to fundamental evolutionary relationships.

Woese was deeply hurt by this general rejection, surviving as a 'scarred revolutionary'. He carried on with his work with a few close colleagues, but kept a distance from the community of biologists thereafter. But he had the last laugh. The ongoing revolution in genomics, over succeeding years, transformed biological understanding, as the technology to read, interpret, and classify genome sequences made vaulting progress. It soon became possible to read the entire genetic code of an organism, and then to sample different environments—soils, seawater, lake floor muds, hot springs—and gather, sort, and classify genetic information from the smorgasbord of organisms (mostly microscopic) mixed among the samples. Woese's three-domain structure of life was soon vindicated, at least to most of the biological community. More archaea were discovered, many being associated with extreme environments: acid, hot, or salty, and so perhaps echoing conditions in which the most primitive life on early Earth lived.

In 2017, a new tree of life was published,[11] based on an assembly of the new genetic cornucopia. It looks more like a starburst than a tree, the many branches radiating out from LUCA at the centre. Each branch is a phylum or a 'supergroup' of phyla. There are 92 phyla of bacteria there: some have recognizable names, like the Cyanobacteria and Actinobacteria, others are surreal-sounding new inventions like the Absconditobacteria, while others are still numbers such as NC10 and TM6. There are 26 phyla of archaea—not as many as the bacteria, but still making up a sizeable chunk of life's tree; among them, to add some belated justice, are the Woesearchaeota (so the man has a whole phylum to himself, and not just a mere species).

Look closely on the bottom right-hand side of the starburst: there, hidden away among a scattering of archaean phyla, are the eukaryotes, our familiar animals, plants, and fungi. At this scale they form a small wedge splitting into a miserly five branches. These have unfamiliar names, for they are new combinations based upon this flood of new information, and so need looking up to work out what they are. There is the Archaeplastida branch, which turns out to be the red and green algae and land plants; and then the Excavata, which are 'protozoans' that include the *Euglena* of school biology classes; and the Chromalveolata, which include brown algae and the tiny silica-shelled diatoms and dinoflagellates (the latter notorious for causing 'red tides' in the oceans); the Amoebozoa, which include not only the amoebae but their larger and more mobile relatives, the slime moulds; and finally, there is the Opisthokonta, which include the fungi and all the animals—the genome evidence indicates that we are more closely related to toadstools than to amoebae).

So, we—and indeed all animals and plants—have been well and truly put in our place within the grand plan of life on Earth. In terms of its diversity at the most fundamental scale, we and indeed all plants and animals are but a tiny part. Most of life's diversity is microbial, and we are only just beginning to realize its enormous extent, or rather, realize the level of our ignorance about it. For most of these bacterial and archaeal phyla, not even one species has been isolated and cultured, so our only acquaintance with them is molecular. The revolution that Woese started has hugely widened, and deepened, the ocean of unplumbed knowledge about life.

This astonishing diversity of microbes is likely an ancient feature of Earth. Woese thought much about the transition from chemistry to life, and the nature of LUCA. For him, LUCA was not so much an individual microbe from which all life can trace its history, as a global microbial community in which the basics of the genetic information that came to characterize all of life on Earth were widely shared. Woese emphasized another phenomenon that was becoming generally recognized late in the twentieth century: horizontal gene transfer.[12] We obtain our genes solely from our parents, and then pass them on to our children, in a 'vertical' pattern. But many bacteria and archaea can exchange parts of their genetic code with each other at any time, either when simply coming into physical contact, or via transfer by viruses that cross-infect them. Such horizontal gene transfer, Woese argued, may have been even more common and probably dominant in the early history of life, and similar chemical exchange processes may well have operated in the preceding, pre-life era too, between entities he termed 'supramolecular aggregates'. It was this phase of communal, not linear, evolution that he saw as one of intense, chaotic experimentation and diversification very early in the history of life, to produce a richness of microbial diversity that persists today.

To Woese, this early cellular evolution could not really have operated on Darwinian principles, because genes were so often transferred horizontally between these imperfect, ramshackle, but inventive life-forms, and not vertically from ancestor to descendant. It was only, he said, when the genetic code became more complex and integrated, and less amenable to such willy-nilly modification and replacement through horizontal gene transfer, that what he called the 'Darwinian threshold' could be crossed. From that point on, it became sensible to think in terms of a tree of life developing on Earth, and evolution could take on the statelier pace that we are familiar with today.

Enter the familiar minority

As our story of life's richness carries on, we move, out of a combination of eukaryotic chauvinism and prokaryotic ignorance, towards the terrain of familiar life that variously possesses characters such as skulls, leaves, and feathers. We mustn't, though, forget the extraordinary new-found diversity of the microbes—and even more the way that the biosphere fundamentally depends on them far more than it does on 'higher' organisms. Remove the eukaryotes from the world, and the prokaryotes will, by and large, carry on as before. Remove the prokaryotes and their many biochemical functions, and the whole of life will come grinding to a halt in days.

In some important ways, the story of the eukaryotes might just as well be regarded as a continuation of prokaryotic evolution by other means, and in rather more sheltered settings. For most of the history of modern biology, key characters of eukaryotic cells—the nucleus, and organelles such as mitochondria and chloroplasts—were thought to have evolved from simple precursor structures, by classical Darwinian gradual evolution. But early in the twentieth century, a few scholars began, more or less independently, to consider a very different, and outlandish idea. They were a diverse bunch in many respects, including the adventurous Andreas Schimper in Germany, the party-loving and lecture-avoiding Ivan Wallin in the USA, and, in Russia, the sinister and depraved Konstantin Mereschkowsky and the societally respectable (at least, he got two Orders of Lenin) Boris Kozo-Polyansky. The idea was simple, and it, and its wider consequences, were neatly articulated by Kozo-Polyansky: organelles within cells were originally free-living bacteria that, sometime in the distant past, were engulfed by another bacterium to create a combined, more complex cell structure—that of the eukaryotes. He called the process 'symbiogenesis' and noted that it was based on co-operation, and not competition—and must happen suddenly, not gradually. The idea was imaginative, bold, and had major implications for evolution—but it was almost universally dismissed, regardless of who had proposed which version of it. By mid-century, it had been virtually forgotten.

Enter a bright, quirky, and determined young biology lecturer in 1960s Boston, USA. Lynn Sagan, later Lynn Margulis, picked up the idea of the symbiotic origin of eukaryotic cells and worked it up more fully for publication, in the context of a half-century's further knowledge. Chloroplasts, mitochondria, and the bases of whip-like flagellae within cells, had all once been free-living bacteria, she declared. She needed all her determination to get the idea even considered. The paper, she later recalled, was rejected some 15 times by different journals, before eventual publication in 1967.[13] This time the idea took hold, and stuck, as evidence (not least the new wave of molecular evidence) piled in to support it. Well within Lynn Margulis's lifetime it became, and has remained, orthodoxy. To her credit, the ebullient Margulis spent a good deal of time and effort making sure that her intellectual predecessors received the credit they deserved, including putting together the first English translation of Kozo-Polyansky's key work.

So—we are all hybrid prokaryotes, and probably more closely related to archaea than to bacteria. When did that step happen? Looking into the fossil record for evidence throws up an array of exasperating—or fascinating, depending on point of view— possibilities, because the criteria are less clear than one might think.[14] Take size, for instance. Most bacteria are tiny, just a few thousands of a millimetre across. A few, though, are giants, like the 'sulphur pearl of Namibia', *Thiomargarita namibiensis*, where each bacterium can reach more than half a millimetre across, its division creating pearl-like chains (from the light-scattering sulphur granules in the cytoplasm) in the sea floor muds off Namibia. And prokaryotes are typically thought of as unicellular, but some can form structures which verge on the multicellular. A few photosynthetic bacteria form chains, in which all the cells are enclosed by an external sheath and in which there is some division of labour, such as between respiration and nitrogen-fixing, and in which therefore some of the cells are dependent upon others for services that they themselves cannot provide. This is effectively multicellularity, albeit at very limited scale. Much more commonly, bacteria form colonies visible to the naked eye, growing by simple division, as on a petri dish. They can also form biofilms: more complex, three-dimensional structures that may include many different species of bacteria and archaea, and that have been described as 'cities for microbes', where there is co-ordination of activity, sharing of nutrients, and sophisticated chemical signalling termed 'quorum sensing'. Some biofilms are very close to us (such as the plaque-forming films on our teeth), while others, that grow on lake and sea floors, are robust enough to be fossilized. In the fossilized forms, where only part of the biological information survives—and rarely indisputable criteria such as nuclei and mitochondria—it is easy to be fooled.

For instance, here and there in strata more than 2 billion years old, is *Grypania*, looking a little like squashed, blackened, and coiled strands of spaghetti. It has been interpreted as an early alga, and therefore eukaryotic—but also as an elongated bacterial colony. And *Horodyskia*, looking like strings of beads in rocks from about 1.5 billion years old, has been compared to a fungus and to a protozoan, but doubts have also been expressed whether it is any kind of eukaryote at all. Searching for the most important transition in life among these fossil scraps has turned out to be a most uncertain business.

Another approach is to look at microscopic single-celled remains, to see if some hint of the kind of complexity and sophistication that comes with the eukaryotic state might be discerned. Recently, a scattering of microfossils were found in rocks almost 2.2 billion years old in northern China.[15] Some were simple forms resembling bacteria. But others had more complex forms, with a layered outer structure, small external spine-like projections, and distinct openings. Resembling the resting stages of marine microscopic algae, these were claimed to be the oldest eukaryotes recognized so far.

So, life on Earth seems to have been exclusively prokaryotic for something between one and two billion years—and, as we have seen, neither dull nor monotonous, but with that astonishing diversity revealed by Carl Woese and his successors. Then, perhaps a little more than 2 billion years ago, some prokaryotes combined to produce the eukaryotic cell. This larger, more complex, more *powerful* entity (fuelled by the energy-providing mitochondria) was the threshold to a different kind of diversity, that afforded by multicellular life. The familiar empire of plants and animals was now made possible.

It was a long time arriving—in any kind of force, at least. Much of the succeeding one and half billion years of the Proterozoic Eon has been dubbed 'the boring billion', partly because the Earth seemed to go through no major climatic changes over that time, and partly because life continued to be dominantly microbial. Eukaryotes were present but, remaining mostly small and soft-bodied, only rarely turn up as fossils. Fossilized chemical 'biomarkers' in the strata indicate, too, that prokaryotes, in the form of those endless microbial mats, and also as microscopic plankton, continued to rule the roost. The cycling of carbon between lithosphere, hydrosphere, atmosphere, and biosphere—that is, the carbon cycle—of those days trundled along monotonously. As a key symptom, the proportion of the 'light' ^{12}C isotope (with six neutrons and six protons in the nucleus) to the 'heavy' ^{13}C isotope (seven neutrons and six protons) in the strata stayed much the same, as the millions of years passed. Climate stayed more or less tepid. There is no sign of any explosive evolutionary radiation of eukaryotes over that time, though they did take a major step more than halfway through this interval.

The first known sexually reproducing organism, given (with due Freudian resonance) the name *Bangiomorpha pubescens*, is found in rock strata just over a billion years old in Canada.[16] It is a red alga, comprising filaments barely a millimetre long and just one cell thick, but fossilized so exquisitely that it preserves both the female 'carpogonia' and the male 'spermatia' that fertilized them, as clearly distinct kinds of cells.

What took the eukaryotic cell so long to invent sex? Perhaps the *effective* refashioning of the eukaryotic cell, it has been suggested, needed so much time, to fine-tune the engulfed and engulfing microbes. It was a combination with something of the character of a pantomime horse initially, one might imagine. Only much later, a properly integrated and functional eukaryotic cell could compete with the long-established and successful prokaryotic model. Nevertheless, sex is an invention that the discoverer of *Bangiomorpha*, the Cambridge-based palaeontologist Nick Butterfield, sees as a key step in the development of complex multicellular organisms. In developing the different kinds of cells that distinguish the male and female organs, sex acts as a kind of test bed for the development of different kinds of tissues generally, while the shuffling of genes that is involved in sexual reproduction provides variation that natural selection can then act on.

That step about a billion years ago ushered in what Butterfield sees as the greatest revolution in life, from a world dominated by unicellular organisms (both prokaryotes and eukaryotes) to one where multicellular organisms took—at least in some obvious respects—the ascendency.[17] As with the evolution of eukaryotes a billion years earlier, it was something of a slow-burning revolution. For 300 million years after *Bangiomorpha*'s origination, the Proterozoic carried on in its humdrum, prokaryote-heavy way. Then, all kinds of hell broke loose.

About 720 million years ago, the Earth was plunged into the depths of an ice age, as a geological time unit given the appropriate name of the Cryogenian Period (the penultimate part of the enormous Proterozoic Eon) began. It waxed and waned over 80 million years, its most intense intervals being the deepest freeze that the Earth has ever experienced, 'Snowball Earth' times that might have seen the entire world covered in a carapace of ice, to resemble one of the frozen moons of Jupiter or

Saturn.[18] Recovery from these glacial pulses was dramatic, the ice realm each time giving way with catastrophic swiftness to seas of tropical warmth. The last major warming marks the beginning of the Ediacaran, 635 million years ago, the last period of the Proterozoic.

Along with these climatic lurches, the carbon cycle, seen as the proportion of 'light' to 'heavy' carbon atoms in the strata, began to show steep switchbacks of enormous size, some the largest in the geological record.[19] A killing time for life, one would have thought, and probably it largely was. But, among the carnage, biological novelties turned up—and some took hold. Just prior to the Snowball times, as the carbon cycle was beginning its first steep plunge, new types of eukaryotic microfossil appeared— little vase-like forms, and also tiny microfossils made of calcium phosphate, the same stuff as our bones and hence the beginnings of real biomineralization. 'Molecular clock' data taken from living organisms—estimates of genetic differences, and the time it may have taken to attain them—suggest that the first poriferans (sponges) may have evolved around that time. The molecular evidence suggests, too, that cnidarians (the group to which corals and jellyfish belong) and the bilaterians (a broad group encompassing most other animals) may have appeared in the depths of Snowball Earth times, though neither have been found as convincing fossils until later, in Ediacaran strata.

Other mysterious eukaryotic fossils turn up in Ediacaran strata. There are tiny 'embryos'—joined clusters of cells looking like a microscopic Rubik's Cube roughly reshaped into spherical form—and most famously, in the later part of this period, the 'Ediacaran biota', the first truly large eukaryotic fossils, some approaching a metre in size. Both are mysterious. Palaeontologists still scratch their heads over the 'embryos' (they are probably not embryos in any true sense) and the 'Ediacaran biota'.

Many Ediacarans have a fractal character, that is, the same body patterns being expressed at large and small scale. They lived largely immobile lives on the sea floor and might have subsisted by somehow extracting nutrients from seawater and sediment; they might include some ancestral forms of our familiar animal groups, or might be some kind of unrelated and ultimately failed experiment in multicellular life, or might include elements of both.

Whatever these bizarre forms may or may not be, they showed that the eukaryotes were finally beginning to challenge the absolute 3-billion-year reign of the prokaryotes. Perhaps this was partly in response to the Earth's dramatic swings in climate and the carbon cycle. Perhaps partly the eukaryotes themselves helped amplify or modulate these environmental switchbacks, by their growing impact on the Earth's chemical cycles. The 'boring' part of Earth's history was certainly over.

The Cambrian explosion, and after

Late in the Ediacaran, life started to move: literally, powerfully, and with a range of consequences that we are just beginning to appreciate. The main culprits here are the bilaterians—those animals with a left and right side, and to go with this they have a front and a back end, and also a 'back' and 'belly' side to their bodies. This encompasses most animals—even worms, if you look at them as scrupulously as Darwin did (a few animal groups are not bilaterian, most familiarly those including the sponges, corals, and jellyfish). The bilaterians had a body structure that, with the addition of some kind of skeleton (which might be the fluid of a hydrostatic skeleton in the case of worms, an external carapace in crustaceans and other arthropods, and internal bones in our case)[20] and muscles, can begin to move forwards, and with purpose.

This movement, initially across the surface of the sea floor, and then down *into* it by burrowing, was the start of a cat-and-mouse game between hunter and hunted that goes on to this day. It also started some pervasive ecosystem engineering, as the early burrowers ripped through the eons-old microbial mats, to allow oxygen to penetrate a little way down into the sediment, to render this shallow subterranean environment yet more attractive for the early, exploratory, bilaterian animals. Meanwhile, the non-mobile, non-bilaterian sponges just above were doing their bit to increase oceanic oxygen levels too, by filter-feeding, cleaning the seawater of suspended organic detritus that would otherwise consume oxygen in decaying. Above them, in the water column, the eukaryotic protists were carrying out a similar function as they preyed on bacteria, packaging up their remains into faecal pellets that fell to, and fertilized, the sea floor.

This concatenation of new biological structures, of physical and chemical changes to the environment, and of new ecological interactions, were likely among the impetus for that extraordinary and unique radiation of animal life-forms that has been called the 'Cambrian explosion'.[21] This event terminated the 3-billion-year long, well-nigh absolute, domination of the world by prokaryotes, and ushered in our familiar world of large, complex animals and plants. The hyper-diverse prokaryotes, as we now know, were not vanquished but simply forced into local retreat and (highly successful) adaptation—we have some 30 trillion bacteria living inside each of our bodies, for instance,[22] roughly the same number as that of our cells—but thankfully occupying a much smaller mass.

The 'explosion' took some 30 million years to accomplish—geologically substantial, but nevertheless representing a real change in pace from the leisurely accumulation of innovation in earlier times. The precise geological time boundary between the Ediacaran and Cambrian Periods (and also the Neoproterozoic and Palaeozoic Eras, and the Proterozoic and Phanerozoic Eons) has been placed at a level where a distinctive corkscrew-shaped kind of burrow appeared in geological strata, dated at 539 million years ago. Nick Butterfield would argue that this was still in reality in the tail end of the Proterozoic world. Some 10 million years later, the radiation of animals seriously began, while the carbon cycle settled back into a somewhat more sedate pattern after the convulsions of the Cryogenian and Ediacaran; the brand-new biosphere really dates from then. By the time the evolutionary radiation abated, several million years later, all the main phyla of animals that we know today, including arthropods (not least the iconic trilobites), molluscs, echinoderms, and even vertebrates, had come into existence.

Amid this inventive, energetic, and combative menagerie, Darwinian evolution could really proceed apace. For us human observers, it becomes clearly evident, by comparison with the murky and mostly impenetrable gloom of the Precambrian record. The relics of these large, impressive, and skeletonized new animals in the suddenly prolific fossil record of the Cambrian and later periods means that we can now follow the twists and turns of evolution, as different species and families of species appear and disappear in the fossil record. Simply being able to recognize individual species and tell them apart from other individual fossil species, with reasonable confidence, just based upon the particularities of their complex physical shapes, is something of a new-found luxury, one we now have in abundance—in these new strata, compared with the meagreness of what went before.

This new cornucopia opens up possibilities such as these— and is a new source of headaches and difficult decisions too.

For instance, with an abundant fossil record, one can begin to measure biological diversity, or biodiversity for short, which is a measure of how many species are present at any one time. Easy enough in principle, but start the work, and problems appear. One can find a set of strata in one place that represents one interval of time, and carefully collect and identify all of the fossils that you can hammer out of those ancient strata. This is a good and practical way to start. At the end of that, you may have a nice haul of molluscs, crustaceans, fish, corals, and other petrified organisms, that you will then classify as best you can, using thick monographs (or, increasingly, their digital equivalent) to help you identify them, into different species, genera, and families (be warned: this is not a task for the faint-hearted: although fascinating, it is *very* time-consuming). To do the job properly, you will have got quite a few friends to join you in the work, as most palaeontologists only have real expertise in one or two fossil groups within one time period (even then, there are typically many thousands of species to become familiar with; it's a lifetime's work). You (and they) will likely have misidentified a few species, missed quite a few others, and of course missed all of those soft-bodied species that weren't fossilized (though, with luck, one or two species new to science will have turned up too).

Add all of this up at the end and one measure of diversity will have emerged, of (part of) the diversity at one time (in one *geological* time interval, that is, which likely represents many thousand years) in one place. Ecologists sometimes call this 'alpha diversity'. Now, walk for a day or two along the hillsides, following, as best you can, that particular stratum representing that slice of ancient geological time (this can be quite a puzzle in itself) until it changes its character, perhaps becoming sandier, or muddier, or giving an indication that currents were of different speed, or the depth of the sea was greater or lesser. As a different kind of rock, it represents a different biological habitat on that petrified sea floor. And so, you (and your friends) go to work again, to repeat the whole process. You will emerge, eventually, exhausted but in proud possession of another long list of fossil species. This list may now have a number of species in common with the first one, especially of those kinds of organisms that could swim or float moderate distances through the sea—but it will lack others from that first locality which did not take to that kind of sea floor, while there will be quite a few different ones too, better adapted to these new conditions. Ecologists may call the difference between these two nearby ancient habitats 'beta diversity', and can recognize exactly similar kinds of change when comparing different kinds of modern sea floor.

Now, the whole party takes a long train journey—something on the scale of the trans-Siberian express—to somewhere that might be a part of that ancient sea floor that was thousands of kilometres away, on the opposite side of the continent, within a different ocean current system and climatic belt. Now it will be quite a trick to pick up that stratal level exactly, because the fossils will be so different. Nevertheless, after much detective work, you find a set of rock strata in these distant parts that look to be the best time match to that original stratum—and, of course, set out to collect, identify, and tabulate all of the fossils again. Now there are likely to be only a few fossil species in common (these help with the time matching), while many of the others will be new, reflecting the different climate and ocean conditions. By itself, this is another local example of alpha diversity, but laboriously put together with all of the fossils from the first two localities, this exercise begins to give a picture of the biological diversity across a broad region—something that is called 'gamma diversity'.

Carry on across the oceans surrounding other continents, where the fossils are different again, and the jigsaw puzzle of ancient biology in time and space of the past is even more difficult, and keep collecting, and collating the species encountered, and a picture of global diversity can emerge, in this one interval of deep time. This is already a huge exercise, and it becomes monstrous when one ventures into other intervals of geological time, to try to track how the Earth's biology has changed through its history—even if it is just the last one-ninth of that history (and, as we shall see, some shortcuts will be necessary at that scale). Nonetheless, this whole mighty exercise was rendered possible— to one newly evolved and inquisitive species of bipedal ape at least—once complex multicellular life evolved to create a new and extravagant kind of fossil record.

Shapes of ocean diversity

Because of the newly abundant fossil record, geological time could suddenly be measured and subdivided much more finely from the Cambrian onwards, using the appearance and disappearance of myriad fossil species as time markers in the precise stratal chronometer that piled up, layer by layer, above the endless temporal wastelands of the Precambrian. Small wonder that 80 per cent of the area of the official geological time chart is taken up with the fine-scale subdivision of this, our most recent eon, the Phanerozoic, which comprises only some 12 per cent of Earth time. A new and more detailed kind of Earth history has become available.

It is not a perfect picture of the biosphere—even of the multicellular world. It shows well what is happening to common animals with hard skeletons, like shells and corals, as these fossilize most easily; even large, rarer animals like dinosaurs and ichthyosaurs have a reasonable fossil record. Small, extremely tough fossils have a good fossil record too, such as pollen in recent geological times, and the spore-cases of microscopic planktonic algae, which range far back into the Precambrian; a single gram of mudstone can contain hundreds of such microfossils. But there are some blank spots in the record—soft-bodied animals, most obviously, but also extremely tiny animals with ultrathin skeletons.

Today, a bewildering variety of these tiny delicate animals inhabit the marine equivalent of soils at the land surface. These are the surface layers of sediment just below the sea floor. Once they became stirred and oxygenated by large burrowing animals—the marine equivalent of Darwin's worms—they could contain a wealth of habitable and nutrient-rich microhabitats, just as in terrestrial soils. And, as in those soils, they contain a marvellous abundance and diversity of microscopic but fully formed animals. Some are shared with those of land soils—the rotifers, for instance, no longer confined to a thin skin of water around sediment grains but having all of the complex inter-grain space to swim through, and the ubiquitous tiny nematode worms. Others are less familiar, such as the ultra-tough tardigrades ('water bears'), the wormlike gastrotrichs ('hairybellies'), the segmented kinorhynchs ('mud dragons'), and the minuscule tentacled cones of the loriciferans. These microscopic faunas are little known today, and their fossil record is very sparse, so much of their important history figures little in the great synthetic overviews of how the tree of multicellular life has evolved and radiated, waxed and waned, over the past half billion years. With such provisos, tens of thousands of laborious studies made by thousands of palaeontologists, working worldwide and piecemeal over the past couple of centuries, can be collated into a history of (multicellular) life on Earth. The overall pattern varies from study to study, and the details are much debated, but some features are clear.

The first hundred million years of this new kind of biosphere was essentially confined to the oceans. The land then was largely barren, greening only a little at its damp, low-lying margins with simple plants such as algae and, later, liverworts and mosses. Even in the oceans, the exuberant multicellular life mostly focused on the shallow seas, with some animal groups taking off to colonize the ocean surface as zooplankton, since large parts of the deep ocean were still prone to being oxygen-starved. In general, it is the animals that catch the eye. Marine plant life seems to have mostly taken the form of short-lived seaweed-like algae and microscopic phytoplankton in the surface ocean, being consumed by the animals as rapidly as it grew. Plants thus formed (and still form) a kind of inverted base to the food chain, with rapid turnover and low biomass at any one time. Complex ocean life of the last half billion years, therefore, has been largely a carnival of the animals.

The beginning of this carnival has its own mysteries. For instance, what if multicellular life was *more* various at the beginning than it is now. This was the suspicion that was growing in the 1960s, when a group of Cambridge-based palaeontologists, Harry Whittington, established professor and unquestioned doyen of trilobite studies, and two gifted young researchers, Derek Briggs and Simon Conway Morris, began a re-examination of a classic Canadian fossil locality, the Burgess Shale of British Columbia, that had been discovered at the beginning of the twentieth century. In these dark shales that date from the middle of the Cambrian Period, and so from the end of the 'explosion', are beautiful silvery fossils. They revealed not just the carapaces of trilobites and other typically armoured Cambrian animals, but the outlines of organisms with delicate, non-mineralized carapaces, and others such as worms that lacked skeletons entirely, complete with traces of eyes, nerves, and guts. It was one of the first, and still among the most celebrated, 'Lagerstätte' to be found. The term, borrowed from German, means 'storage place', and that is exactly what it is, for the bodies of ancient organisms: an exceptional window into the past.

The weird and wonderful menagerie of Burgess Shale fossils, many of them new to science, had originally been shoehorned, one way or another, into the familiar phyla of today—this is what the early twentieth-century geologists knew, after all. Then, half a century later, came Whittington's team for a closer look. They found oddities that they could not fit into the usual categories. Among them was *Opabinia*, a long-snouted creature with five eyes, and *Hallucigenia*, a beast that seemed to walk on stilts, with a row of tentacle-like structures along its back, and others that seemed not to fit the usual categories of life. Perhaps, they said, these are representatives of new phyla, entirely different major branches of animal life, that have since become extinct. These suggestions were made in the discussion sections of a succession of long, highly technical academic papers. Palaeontologists pricked up their ears, but there was little wider awareness that some sort of revolution might be brewing.

Enter a brilliant, adventurously minded Harvard palaeontologist, Stephen Jay Gould. Already famous outside his field for books of beautifully written essays on evolution, where such things as baseball statistics and operatic arias were brought into the narrative, he was also something of an *enfant terrible* of the science—both a great public defender of Darwinian evolution and a man determined to change widely-held perceptions about the process. He took the accumulated work of the Whittington team—a set of difficult, dry, academic works—and turned them into a popular science book, *Wonderful Life*. It became a bestseller. Not only that: rather like one of the eighteenth-century savants, he used this public-focused book as a vehicle to bring new ideas into scientific debate. The Cambrian explosion, he declared, had produced a *wider* range of fundamentally differ-ent animal body plans than we see today; this is called a greater 'disparity' (as opposed to 'diversity', which is a measure of the number of species). Some of these body plans survived through geological time until today, and some did not, and the reason was as much chance (something he called 'contingency') as it was ecological fitness. Rewind the tape of time from the same starting point, he said, and there would likely be a very different set of survivors, to give a very different kind of biology today—perhaps one dominated by creatures with five eyes and walking on stilts.

The idea made a splash, as did many of Gould's intellectual challenges. Could life today really be more impoverished, as re-gards the variety of major body plans, than life way back in the distant Cambrian? There was immediate pushback from the palaeontological community—a reaction that clearly showed how provocative ideas (*well-constructed* provocative ideas, that is) are a marvellous stimulus to one's colleagues to go back to re-examine the evidence. Further study of the proposed weird new phyla showed them to be no less weird, but perhaps not quite as new and different as regards fundamental animal relationships as Gould or the Whittington team had suggested. Another palaeon-tologist, cleaning some of the mudrock matrix away from a *Hallucigenia* tentacle, found a claw at its end—so it was a foot, and not a tentacle! And the 'stilt-legs', once turned upside down, were revealed to have been fearsome protective spines on what turned out to be a 'lobopod', an animal related to the obscure onychophorans (velvet worms)[23] that still live today amid the leaf mould of tropical forests. The bizarre *Opabinia* was reclassified among the beginnings of the known arthropod lineage, and some other of the 'new' phyla were plausibly reeled in to sit among the roots of the old ones.

Mysteries remain, such as *Wiwaxia*, an enigmatic tangle of armour and spines that might be a worm or a mollusk (or neither). But it is now generally thought that the tree of animal life has endured more consistently since its beginnings than in Gould's striking concept. And, as a philosophical rider to that, Simon Conway Morris, one of the original wellsprings of this intellectual adventure, opined that if the tape of animal life was replayed again from its beginnings, it would likely lead to similar general patterns to the ones we see today.[24] It's an observation that, if true, might apply to habitable planets beyond Earth too.

After those riotous beginnings, what happened to life?

Ocean designs

It was Jack Sepkoski, one of Stephen Jay Gould's students, who showed a new picture of life's grand patterns over the past half-billion years. This was no coincidence: they were both part—and instigators—of a conscious attempt to transform their science, to give it real meaning beyond the ranks of fossil enthusiasts and geologists. High-profile, highly mathematical efforts to unlock the secrets of evolution were at the time, in the mid-twentieth century, largely being spearheaded by the geneticists. They had a low opinion of palaeontologists, thinking of them as little more than stamp collectors. One of their leading figures, the English geneticist John Maynard Smith, expressed it pithily, '[T]he attitude of population geneticists to any palaeontologist rash enough to offer a contribution to evolutionary theory has been to tell him to go away and find another fossil, and not to bother the grownups.' Gould and his colleagues were determined to change this attitude—and they succeeded.[25]

Sepkoski's primary field area was not some fossil-strewn cliff (though he did a little of that kind of work, too). It was a library (an old invention) and a computer (then, a new one). He was building a database (also something of a new concept at the time) in order to seek, if not quite the meaning of life, at least its grander patterns. The library held the data: many thousands of individual studies of the 'stamp collectors'—the geologists and palaeontologists who, over a century and more had slowly, patiently, arduously dug the fossils out of strata from around the world and identified, dated, and catalogued them. Sepkoski patiently began amassing their results, to catalogue when these different types of prehistoric animals lived and died, and designed computer programs (a very new skill, in which he soon surpassed Gould) that would put all this together into a narrative of life on Earth.

Sepkoski's iconic narrative takes the form of a graph, now as familiar to most palaeontologists as is a *Tyrannosaurus rex* skeleton. Conjured out of the depths of his primitive computer, it shows a line rising and falling—though overall rising—across a half-billion-year span. The line shows the diversity of life on Earth. Its units are not species (there were far too many of these for even a man as assiduous as Sepkoski to count) or even the clusters of related species that form a genus (as in the genus *Homo*, in which a number of species have been recognized, including *Homo heidelbergensis* and *Homo neanderthalensis*, and our own species, *Homo sapiens*). Sepkoski found it most effective to use the clusters of related genera (the plural of genus) that form families: our own family, for example, is the Hominidae. And he stuck to the marine record of life, as the record of life on land seemed too patchy and difficult to make sense of. At this scale, and in this realm, he was able to build up a vista of life's fortunes.

The line stutters upwards in the Cambrian Period, which, after the 'explosion', might be seen as a kind of bumpy, trial-and-error apprenticeship for multicellular life, ending at some 180 families. It put in place what Sepkoski called the 'Cambrian fauna', dominated by such familiar fossils as the trilobites, together with less well-known ones such as the hyolithids (animals that lived in a cone-shaped shell topped with a couple of Viking-like 'horns') and monoplacophorans. The monoplacophorans proved more spectacular as regards longevity than the coelacanth—they were thought to have become extinct 400 million years ago, until living examples were dredged up from the sea floor off the coast of Mexico in 1952.

Then, in the Ordovician Period, came a great 40-million-year expansion of life forms—an evolutionary radiation—that roughly tripled the number of families, to nearly 600. The new families represented the arrival of new kinds of animals—cephalopods, corals, starfish, sea-lilies, brachiopods ('lampshells'), and others—that were to dominate marine life for the next 200 million years, the remainder of the Palaeozoic Era, and so termed by Sepkoski the 'Palaeozoic fauna'. Many of these families are still with us today in substantial numbers. The Cambrian fauna, meanwhile, began a slow decline and—bar a few rare and lonely individuals such as that elusive Mexican monoplacophoran—has disappeared. But here's a curious thing. For most of that 200 million years, Sepkoski's Palaeozoic line is near-horizontal: individual families came and went, but their total number stayed more or less the same. The implication is that some kind of carrying capacity had been reached in the oceans, that would support only so many different kinds of organisms. There were a couple of sudden dips that represented mass extinction events—one at the end of the Ordovician Period, about 440 million years ago, and another near the end of the Devonian Period, about 360 million years ago. But, biodiversity soon (that is, over several million years) recovered to stay on the same eerily flat plateau.

Then a crash, sharply visible even at that scale: the 'great dying' at the end of the Permian, 250 million years ago, a carnage which halved the number of families—and as a consequence slashed the number of species by an estimated 95 per cent, because to kill off a family, it is necessary to kill off *all* of its members. Recovery was slow, over tens of millions of years, but eventually brought with it a new pattern, of Sepkoski's third, great, 'Mesozoic fauna'. Dominated by molluscs, fish, crustaceans, sea urchins, and marine reptiles and mammals, this fauna carried on into the present Cenozoic Era, and is still essentially the kind of life we have in the sea today.

But life's pattern in the Mesozoic was different. The recovery continued as new families evolved, and carried on, and on. The number of families rose and rose, a trend halted only briefly by two more mass extinction events, one at the end of the Triassic Period, 200 million years ago, and the more famous end-Cretaceous event, 66 million years ago. The number of families reached that Palaeozoic ceiling of near 600 early in the Cretaceous Period, and then kept going. Even *after* the end-Cretaceous catastrophe, the number of families was greater than at any time in the Palaeozoic. After the subsequent recovery, the rise continued; for present times, Sepkoski's database counted over 900 families with the curve still rising. So now there seemed to be no clear limit to Earth's diversity.

These mysteriously contrasting patterns have given rise to arguments about the Earth's carrying capacity for life that persist today. Sepkoski's patterns have since been puzzled over, reworked, and reshaped many times, and the kinds of databases he bequeathed have grown enormous. The importance of such studies to pondering the grand questions of planetary life are now unquestioned: Gould, Sepkoski, and their fellow conspirators had brought palaeontology to the high table of evolutionary science. But there were still surprises to come. Life on land, for instance, followed a different tune.

The biosphere moves on to the land

Jack Sepkoski's diagram has acquired such iconic status that one can forget that it is an image constructed just of marine life, and from there it is all too easy to make the inference that its patterns represent those of all multicellular life on Earth. Abundant, complex life on land came late, after all. The terrestrial invasion only began properly mid-way through the Palaeozoic Era, 150 million years or so after the 'Cambrian explosion', as the mighty, amphibian-haunted forests of the Carboniferous Period spread widely across the Earth. Small wonder: the land is not only dry but subject to much larger swings of temperature than the sea, while gravity is a cruel force to overcome for organisms used to the supportive medium of water. The evolutionary re-engineering that was needed to survive in this harsh terrestrial setting was substantial, for both plants and animals. One might imagine this testing environment always taking a back seat, in evolutionary terms, to the wide and deep ocean. And even if one admitted that the terrestrial fossils may be able to tell an important story at this scale, an additional hurdle remains. These pioneering land organisms not only needed to have lived long enough to tell their story, they had to have died appropriately too, in such a way as to preserve their remains for the many millions of years for future human palaeontologists to unearth and interpret. Now, one of the rules of thumb in geology is that the sea is an environment of sedimentation, burial, and fossilization, while the land is one of erosion—and therefore of the destruction of fossil evidence. Exploring the history of life on land in the way that Sepkoski interrogated the marine strata seemed a tall order.

A successor in the kind of deeply questioning, data-rich palaeontology that Gould, Sepkoski, and others set in train is the genial British scientist Mike Benton, who has an equally enviable knack for asking the right questions—and the patience to amass the evidence to answer them. First, he made the point that, on closer examination, and with enough patient searching, the poor fossil record of land-based organisms turns out to be one of the less well-founded rules of thumb. There are good burial places on land, for both plants and animals: river valleys, lakes, large deltas—such as those which buried the immense Carboniferous coal forests that we now excavate and burn so freely, for instance. All of these can yield fossils, often abundantly. What do those fossils reveal?

The land, says, Benton, dances to quite a different tune from the sea.[26] It must, because there are two very large dots to connect. If we look at the biodiversity today, most species live on land, even though it occupies only a third of the Earth's surface. There may be as many as 25 times more species on land than in the sea. And, as complex, multicellular life had less time to evolve on land (not much more than 400 million years) than in the sea (about 600 million years), it must have diversified far more quickly.

Benton compiled for land-based fossils—both animals and plants[27]—the same kind of database that Sepkoski had used so eloquently for animals in the sea. The line now slowly rose from mid-Palaeozoic times, then jumped a little in mid-Carboniferous time, when those coal forests grew. There followed a little plateau, until the end of the Palaeozoic, of some 200 land-based families, so roughly a third as many as were present in the sea then. The 'great dying' of the Permo-Triassic mass extinction saw a modest dip in families, and from then on an increase in diversity that, despite minor (at this scale) wobbles at mass extinction events, seems to have picked up speed all the way through the subsequent 250 million years, with just possibly a levelling off in the past few tens of millions of years.

Today, there are something like 1,500 families of organisms on land; that's not many more than those in the sea (about 1,200, including plants)—but land-based families seem to be much more species-rich than those in the sea, to explain that hugely greater difference in total species diversity. And in turn, those land-based families belong to fewer of the major biological groups, the phyla: in the sea, there are some 43 phyla, but only 28 of those made it onto land (animals were particularly shy of land, with only 12 out of the 32 marine phyla making the transition).

Patterns of life on land, thus, differ markedly from those in the sea (it almost seems like a different planet). There is less diversity at the level of major groups, far higher species diversity, and a pattern through time that shows far less of a sense of 'equilibrium', of the land having reached some kind of carrying capacity for species numbers. What could be going on?

One difference lies in how plants and animals relate. By contrast with the plant-poor 'inverted pyramid' of the sea, the pyramid of life on land rests firmly on a plant base, with—before human intervention—forests and grasslands stretching over virtually all the land that is not ice, desert, or high mountain. Most, by far, of the Earth's total biomass (land and sea) is trees. This provides both a permanent source of food for animals, and also myriad forms of shelter and microhabitat, from that prolific meiofauna-rich soil to the top of the leaf canopy. Life, in this way, engineers its own surroundings and, as it evolved, created yet more kinds of space and opportunities for life, with the circulation of nutrients always aided and abetted by the ubiquitous microbes and the fungi. In the oceans, this kind of intricate, three-dimensional habitat is really only provided by the coral reefs, which occupy only a small fraction of ocean habitat, and that mainly in the low latitudes. One might think of the land as being largely covered in wraparound coral reef, as regards the potential for creating niches for new species to appear and diversify.

It's heaven for life—and one that, in that half-billion-year overview, still seems to be finding more space for new kinds of organisms. As Mike Benton noted, '[T]he idea of unconstrained expansion of life . . . is somewhat bewildering.' Presumably, there are limits somewhere, set by the finite amount of mass and energy at our planet's surface. And our slowly brightening Sun means that life on Earth as a whole has perhaps just a billion years still to run, before the tightening greenhouse effect snuffs it out. But in the meantime, a view of our planet that implies new perspectives of grandeur is worth holding on to, as we consider how humans are changing the numbers, and the narrative.

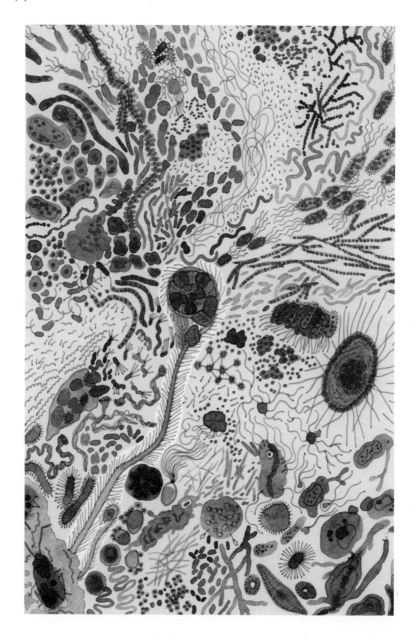

3

THE PLANET IN A GARDEN

The Welsh village of Llanbadoc, in Monmouthshire, today has some 800 inhabitants, a sawmill, a college, a church, and a prison. Two centuries ago, one would guess that it probably had rather fewer in the way of both people and amenities. A place far too insignificant, one might surmise, to illuminate the grand designs of life on this planet. Nevertheless, in 1823 it was birthplace to Alfred Russel Wallace, who shed a good deal of light on those designs.

His beginnings were unpromising. Alfred was born to English parents, who, during his schooldays, fell on hard times. The 13-year-old was forced to leave grammar school and learn a trade. After briefly living in London, he joined his oldest brother William to work as a landscape surveyor in the Welsh Borders. In 1843, at the age of 20, he moved to the Collegiate School in Leicester to teach map-making and surveying. There, his fortunes changed. He met Leicester-born Henry Bates, and an adventure began that would take both men across the world and spark important insights into the fundamental patterns of life on Earth.

Wallace struggled with finances throughout his life and so did not fit the mould of the typical Victorian gentlemen scientist, making even more extraordinary the endeavours that took him into the jungles of the Amazon and South East Asia. This is in sharp contrast to his more famous contemporary, Charles Darwin, who was financially secure through the industrial legacy of his grandfather, Josiah Wedgwood. Nevertheless, like Darwin, Wallace assembled his knowledge of biology through extensive travelling and by making prodigious collections of specimens. This interest seems to have been sparked by Henry Bates, who at the age of 19 had published a short article in the journal *The Zoologist* on beetles from damp places in Leicester and the surrounding hills.[1] Grubbing around in ditches, Bates already displayed the keen eye of a naturalist, observing seasonal changes to insect faunas. This interest soon infected Wallace too, who for the rest of his life would be an enthusiastic collector of beetles.

Bates and Wallace avidly read many of the influential natural history texts of the first half of the nineteenth century, and after Wallace left Leicester in 1845 to look after the business of a brother who had died, they kept up an active dialogue in letters. One of the books they read was *A Voyage up the River Amazon*, by William Henry Edwards, an American businessman and natural history enthusiast, who went on to write a classic monograph of North American butterflies. Enticingly written and conjuring images of exotic and beautiful landscapes, plants and animals, *A Voyage up the River Amazon* was a key inspiration.[2] The two men determined to follow in Edwards' footsteps.

After hatching their plans for an Amazonian expedition in 1847, Wallace and Bates left the port of Liverpool in April 1848, travelling on a tall ship destined for South America. They arrived at the Brazilian coastal city of Belem in late May. After a year acclimatizing and collecting biological specimens around the city, the two men separated. Bates was to spend the next 11 years in the Amazon basin, amassing an extensive collection of insects, and only returning to England in 1859. He was to become a famous naturalist in his own right through his work on mimicry in butterflies (now called 'Batesian mimicry' in his honour), a widespread phenomenon in which a harmless species copies the colour patterns of a toxic species as a deterrent to predators.

Wallace meanwhile made for the Black River (Rio Negro), a tributary of the Amazon, so called because its waters are tea-coloured from masses of decaying plant matter. And, though being only a tributary of the mighty Amazon, it is itself one of the largest rivers in the world, equivalent in scale to the Mekong River of East and South East Asia. Wallace's expedition along the Black River was to end, though, in disaster. The ship on which he was travelling to England in 1852 caught fire on 6 August, taking all his carefully collected specimens and most of his notes to the bottom of the sea. Wallace himself had a narrow escape, spending 10 days aboard a small rowing boat drifting in the Atlantic before being rescued by a passing Norwegian ship en route to London. Despite this loss, Wallace managed to publish a number of articles on his discoveries in Amazonia. And, unperturbed by the misadventure, in 1854, he set off on more travels, this time to South East Asia. This expedition was to last until 1862.

Wallace's travels in South East Asia were prodigious. With little by way of formal funding, and often living rough in a physically challenging climate, he covered over 22,000 kilometres from the Malay Peninsula through the Indonesian archipelago to Papua New Guinea. Wallace collected over 100,000 specimens of his beloved insects, as well as many thousands of bird specimens.[3] His collections would help to build the foundations of the science of biogeography, the study of the patterns of distribution of plants, animals, fungi, and microbes across the world. They would also lay the foundation for his own insights into biological evolution.

The revelation that organisms evolve by adapting to their environment appears to have come to Wallace during a bout of fever in early 1858.[4] His ideas on natural selection and the origin of species were developed quite independently from Darwin. Nevertheless, he was aware of Darwin's interest in the theme, and wrote to the older man outlining his ideas. It was one of those key moments in science, making Darwin realize with a shock that his long delay and many hesitations about publishing his voluminous evidence that species were not forever fixed but 'mutable' ('it is like confessing a murder', he wrote) meant that he, and the central part of his life's work, were in danger of being scooped.

The catalyst acted quickly. Friends of Darwin assuaged his despair by suggesting that the two men collaborate, which Wallace agreed to do. They published their revolutionary idea together on 1 July 1858 at the Linnaean Society in London, as a presentation 'On the Tendency of Species to form Varieties'. Neither man was present. Darwin was ill, and also grieving over the death of a son, while Wallace was in the midst of his travels, on the Moluccan Islands in Indonesia. The presentation of their joint work was marked by silence, with no discussion: the audience perhaps sensed that the subject was too new—and too ominous. If so, this awareness was not shared by the then President of the Society, Thomas Bell. Reviewing the year, he said, 'The year which has passed ... has not, indeed, been marked by any of those striking discoveries which at once revolutionise, so to speak, the department of science on which they bear.' One wonders if he reconsidered his view, when Darwin's famous book *On the Origin of Species* was published the following year, and controversy over the theory of evolution was definitively ignited.

Being part of such a major revolution in scientific thought should have been enough for any one person, it might have been thought. But Wallace was puzzling over other big ideas. Crossing from Java and Bali into the small East Indonesian island of Lombok he wrote in his journal, 'Plenty of new birds ... Australian forms appear. These do not pass further West to Baly & Java & many Javaneese birds are found in Baly but do not reach here.' Wallace may have relied on some local knowledge to make this judgement, having at that time not travelled farther east, and having never visited Australia. But he also carried with him a compendium of birds by the French naturalist and nephew of the Emperor Napoleon, Charles Lucien Bonaparte, and his keen eye would have spotted the absence of those birds characteristic of the islands of Java and Bali to the west.[5] Wallace had thus observed and recorded the 'Wallace Line', though it only became known by that name some years later.

The Wallace Line is one of the most fundamental divisions between the animals and plants on Earth. It demarcates the faunas of South East Asia with their monkeys and apes, from those of Australasia with their kangaroos and wombats. These strikingly different faunas were separated by an impassable deep-sea barrier—the Lombok Strait—that cuts between the small islands of Bali and Lombok and runs on northwards between the larger islands of Borneo and Sulawesi. This line is a physically small but important part of the mosaic of animal, plant, fungal, and microbial biogeographical patterns identifiable across the Earth, which are controlled by climate, geography, and oceanography.

While Wallace was scratching around in the jungles of South East Asia—and Bates was still amassing evidence for insect mimicry in Amazonia—another English biologist, Philip Lutley Sclater, was establishing the global distribution of the world's birds from the relative safety of published accounts. As a young boy, Sclater had already developed a strong interest in ornithology, and this became his life's work. He published many important works on natural history, amassing over 1,400 in all, many with lavish illustrations of both birds and mammals, but his most famous work laid the foundation for the modern science of biogeography.

In a study published in 1858, and compiled before evolution had yet entered the arena of discussion, Sclater identified six areas of 'creation', neatly outlining the distinction between the birds of North and South America, and drawing the line cleanly through the 'Table-land' of Mexico. Sclater's work was remarkably prescient and, almost willing someone to find the 'Wallace Line', he wrote that it is 'not yet possible to decide where the line runs which divided Indian zoology from the Australian'.

The geographical patterns of animals identified by Sclater, when assembled on a map, form a jigsaw puzzle of interlocking land regions. Eight are now recognized, of which the Australasian region, east of the Wallace Line, is immediately obvious from its geographical isolation by seaways. The biggest of these regions, called the Palearctic, is more enigmatic, as it runs through landmasses from Iceland, through Europe and North Africa, on into Central Asia and East Asia, to finish in Japan. Its demarcation is a combination of marine barriers on its eastern and western margins, and topographical and climatic barriers to the south, though exactly where the southern boundary is drawn in the deserts of North Africa and Saudi Arabia remains a matter of scientific debate. Sclater had already identified the Palearctic when he wrote, 'Europe and Northern Asia are in fact quite inseparable. So far as we are acquainted with the ornithology of Japan—the eastern extremity of the temperate portion of the great continent, we find no striking differences from the European Avi-fauna.' Wallace later recorded these major patterns in his book *The Geographical Distribution of Animals*, but it was Sclater who first discerned them.

Within this jigsaw pattern of the major regions of life on Earth, animal and plant communities also respond to the patterns of rainfall and temperature. And so, the Palearctic includes, from north to south, large areas of tundra, conifer and deciduous forest, grasslands, and the deserts of North Africa, the Arabian Peninsula, and Central Asia. Southwards into the Neotropic, Afrotropic, and Indo-Malay regions, extensive areas of rainforest grow. Here and there mountain ranges burst through these patterns, and grow temperate forests at tropical latitudes, as Humboldt and Bonpland had observed on their journeys through the forests of equatorial America. These climatically controlled patterns are called ecoregions, and any one biogeographical realm, like the Nearctic of North America, will contain many of them.

Earth's terrestrial biogeographic patterns have been a key feature of life on Earth for hundreds of millions of years, changing in their shape and size as continents drift apart and then collide. But changes are now afoot that, uniquely in Earth history, are affecting every part of the globe more or less simultaneously. To examine them, we will travel across some far-flung regions in these pages. But it is easiest just to start in the garden.

The planet in your garden

The horticulturalist Robert Hart was a pioneer, but one who acknowledged he was standing on the shoulders of a body of human knowledge stretching deep into prehistory. His temperate forest garden, built on the Silurian rocks of Wenlock Edge in the Shropshire hills of England, was constructed as a source of food, a natural and self-sustaining ecosystem, and a place where he and his brother could find well-being and connect with nature.

Hart organized his garden from observing the structures of naturally growing forests, using six levels from the tree canopy at the top, to the soil biota of fungi, roots, and rhizomes at the bottom, with herbaceous and shrub layers in between, and connected by a seventh component of vertically creeping plants. In the foreword to Patrick Whitefield's book[6] *How to Make a Forest Garden*, Hart wrote passionately about our deep connection with woodlands, which in prehistoric times provided our ancestors with food, shelter, and clothing.

What kinds of plants, then, might have surrounded a prehistoric community in Shropshire—one that might have existed 10 thousand years ago say, not long after the end of the last Ice Age? There would have been such trees as oak, hornbeam, aspen, beech, yew, hazel, lime, willow, and blackthorn in the forests that clothed the ground after the ice retreated. Clearings may have been lined with juniper, buckthorn, dog rose, and sweet briar. Our ancestors may, perhaps, have delighted in flowers such as primrose, wood anemone, lily of the valley, Pasque flower, and hellebore, cursed the abundant flying, walking, crawling, and stinging insects, hunted the wild boar and deer, feared the wolf and auroch, and kept a safe distance from the lynx. If they ventured out at night, they might have happened upon the eerie lights of the will-o'-the-wisp dancing above the sodden ground, that led them to weave legends of lost souls and ghosts that tempt unwary travellers to their doom.

Fast forward 10 millennia, and the gardens of most homes consist of straight lines, shaven lawns, concrete patios, and neatly trimmed borders, microcosms of how we try to control nature in our cities. The ramshackle corners of our gardens hide behind the bins, or at the rear of the garage, or behind the compost heap. In such places a remnant of the natural ecology might try to get a temporary toehold. A bluebell or buttercup might be tolerated. Occasionally a weed, like a dandelion, will poke its head through the grass, only to be clipped to its roots by the mower, or shrivelled by the spray of the weed killer. The plants in such a modern garden would be mostly bewilderingly unfamiliar to our ancestors. The Japanese Sakura, say, growing in one corner, with its beautiful pink blossom in March. A Chinese rhododendron, dwarf or full size, in another corner, and perhaps a *Dicksonia* tree fern from the temperate rainforests of New Zealand.

The biodiversity of these gardens is high, if measured by individual species alone. And, the effect is felt well beyond the gardens. In the UK alone the Royal Horticultural Society suggests there are about 1,400 species of introduced plants[7]—roughly equivalent to the number of native plant species—that have escaped the confines of gardens to live freely, part of the new biological landscape. In ecological parlance, they have become 'naturalized'.

Of these introduced plant species, a little more than a hundred have become invasive. Either because they have arrived without the natural enemies of their native land, or because they spread very quickly, or simply because they are superbly efficient at taking over local resources, they have pushed native species aside and come to dominate local ecosystems. Some of these newcomers are already well known to gardeners and homeowners, like the Japanese knotweed—unproblematic in its native Japan but a highly expensive problem in the UK, the giant hogweed, originally from the Caucasus, and the rhododendron spreading relentlessly across hillsides. Five introduced water plant species have had such a severe environmental impact that they have been banned from sale, including the Australian swamp stonecrop, which has a devastating effect on the ponds it invades, strangling the life out of them by forming a dense green mat of vegetation at the surface.

Some of the longer-established species already seem to be traditional parts of gardens and the countryside. In Britain, the horse chestnut tree, the source of conkers for generations of schoolchildren in times before electronic games, was introduced from its native Greece in 1616, while the sycamore, with its characteristic winged seeds, arrived only a little later from central Europe and Asia. In the branches of these trees one might see collared doves from Asia and grey squirrels from North America. Hopping around on the ground there may well be rabbits that were brought from their native France and Spain to Britain by the Romans, but only really became established in the wild in the twelfth century. They have famously spread more widely, being introduced to Australia in the late eighteenth and nineteenth centuries for food and sport, before escaping and growing to populations of hundreds of millions that, despite the desperate efforts by humans to shoot, trap, and poison them, have nibbled their way across the continent.

Many exotic species lie hidden. In the soil, earthworms are becoming less common in growing numbers of places, preyed upon by invasive flatworms from Australia and New Zealand which have few of the soil-conditioning properties of their victims. And these are only the larger, more obvious, well-studied species of the extraordinarily complex soil ecosystem. Changes to the smaller, cryptic yet (originally) highly diverse meso- and microfauna are poorly known, let alone those of fungal and microbial soil communities. Among the casualties, though, are the ghostly lights of the will-o'-the-wisp.[8] These eerie flames are now hardly ever sighted, probably because those wide, swampy wildernesses have now been drained, and converted into tidy plots of agricultural lands and gardens. As the microbial communities were transformed, the will-o'-the-wisp flickered out.

Since his death, Robert Hart's garden at Highwood Hill on Wenlock Edge has lain fallow for two decades.[9] It is slowly returning to a more natural state, the elevating tree canopy now blocking out much of the light to the forest floor, though here and there a few shrubs near ground level still grow. This is a fitting end to a forest garden, that now becomes a natural forest in its own right. Elsewhere, we humans have profoundly altered the ecosystems around us by introducing new animals and plants. But animal and plant populations have mixed, too, before humans appeared on Earth.

The Great American Interchange

South America and Australia are currently separated by an expanse of about 7,000 kilometres of Pacific Ocean, which has acted as a highly effective barrier to species migration. The native faunas of these two continents clearly show their separateness, exemplified by the llamas and capybaras of the former and the kangaroos and duck-billed platypuses of the latter.

And yet surprisingly, some of the plants and animals that survive in the forests of South America and Australia—biogeographical regions that are now far apart—retain a distant relationship with each other, through common ancestors that lived more than 100 million years ago, when these two continents were joined together in one gigantic landmass called Gondwana. The different pieces of that leviathan were broken up by the actions of continental drift, driven by the extraordinary planetary mechanism that is plate tectonics.[10] It was Philip

Lutley Sclater—the man who discerned the geographical patterns of birds—who was also the first to observe these remnant
relationships when, in 1864, he published *The Mammals of Madagascar*, noting that lemurs occurred in India and Madagascar,
but not in Africa, and suggesting that long ago Madagascar
and India must have been joined in a single continent which
he called Lemuria. Although he was working long before the
theory of plate tectonics was developed, Sclater had observed
a phenomenon that is now well known to geologists: that, as
previously conjoined continents drift apart, their faunas and
floras retain some similarities, even though the individual plants
and animals gradually change, as evolution in areas that have
become geographically and climatically isolated takes separate
paths. And, conversely, when continents are brought together,
whole ecologies may change, as invaders arrive across land
bridges to conquer new lands.

As they drifted apart for millions of years, the faunas and
floras of South America and Australia retained a memory of
each other, in monkey puzzle trees, lungfish, and marsupials.
Antarctica—another part of Gondwana—drifted too, remaining an island continent, but moving southwards to keep its
lonely sojourn over the South Pole, whilst its land-based flora
and fauna simply froze to death. Other parts of Gondwana—
Africa, Saudi Arabia, and India—broke away to drift northwards,
eventually colliding with Eurasia. For more than 100 million
years South America drifted in near isolation, evolving many
remarkable creatures like anteaters and armadillos. In its past
there were also elephant-sized ground sloths, gigantic crocodiles,
and giant flightless birds—the 'terror birds'—that were ferocious
meat-eaters. As Africa and South America remained close for
a time, before the South Atlantic fully opened, some animals

managed to island-hop to become the ancestors of the capybara and South America's monkeys. Some may have made this journey on driftwood, and these natural invaders diversified in their new landscape of South America. But for all this time there was no connection with North America.

The two continents, North and South America, finally began to approach each other about 10 million years ago, and the appearance of the first American invaders, north and south, begins shortly after that. Before the two continents were bridged by an isthmus, rodents and skunks had already travelled south, and some sloths had made the passage north. Thereafter, from a little less than 3 million years ago, the Great American Interchange took place. Horses, deer, bears, sabre-toothed cats, and cougars headed south. Sloths, armadillos, and even one species of terror bird headed north. The invasion of South America by North American cats led to the demise of many of its indigenous large predators, but some groups, like ground sloths, fared better. Although South American faunas witnessed widespread extinction, none such occurred in the north. Nevertheless, many migrants from the south fared well in the north, and some extended their ranges as far as Canada.

In Australia and New Zealand, where a land bridge to another continent was never formed, their plants and animals, especially their marsupial mammals, continued to thrive in splendid isolation. So too did the inhabitants of oceanic islands far removed from continents, such as Mauritius and Hawaii. But even here all was about to change, though not at the hands of continental drift. The jigsaw puzzle of life, the patterns of biogeography controlled by the position of the continents and oceans that had existed for countless millennia, would begin to break down as the alien invaders arrived.

Alien invaders

Humans have been refashioning the ages-old geographical ranges of animals and plants since the beginning of domestication, some 14,000 years ago or more. This likely began with 'man's best friend', the dog, and evidence is found of such early canine domestication in many places from Kamchatka to Europe. Still earlier, some 23,000 years ago, hunter-gatherer peoples living in a tiny village between the Mediterranean Sea and the Sea of Galilee in Israel were already consuming the wild forerunners of barley, wheat, lentils, peas, and oats, and they may have been practising a rudimentary type of farming.[11] Here people were living in small huts made from willow, oak, and tamarisk, and artefacts within the huts include wooden objects, beads, flint and stone tools, and animal bones and shells. Alongside them in this more settled life were those ubiquitous cohabiters of humans, rats.

Beyond the shores of the Sea of Galilee, other animals and plants would become domesticated and subsequently spread across the world. Crops like maize, that were first cultivated in Mesoamerica 9,000 years ago, have become global in their distribution, as have animals like turkeys and chickens, which are eaten in huge numbers each year. Wherever geologists look, these changes to plants and animals have left a fossil signature, identifiable—to take just one example—in the sediments of cores sunk into the bottom of a Kenyan rift valley lake, that record many centuries of history. These cores show maize pollen appearing about a metre below the lakebed surface, marking its arrival in this area during the seventeenth century. And above the maize pollen, the pollen of pine trees appear in the cores, with their distinctive shapes that mimic the head and ears of the Disney character Mickey Mouse, and indicate the introduction of pine trees to this region by European colonists in the early twentieth century.[12]

There are thousands of alien species globally. The rate of their introduction has been increasing over the past few centuries, and further accelerated since the mid-twentieth century.[13] These species range from bees, frogs, rabbits, and snails, to oil palm and acacia, and all continents are affected, with the exception of the Antarctic landmass. Some of these aliens have crept in by stealth, and others by deliberate introduction. They have proliferated in landscapes to the extent of entirely changing ecosystems. Many are also remarkably good at proliferating in environments that have suffered environmental damage at the hands of humans.

The introduction of new species also threatens to upset the delicate balance of the oceans. Here species are sometimes moved deliberately for aquaculture, like oysters, but many have been moved around the planet invisibly and unwittingly, in the ballast tanks of ocean-going ships. Some of these marine invaders have had a serious impact on local ecologies, like the Asian lionfish, which has taken up residence along the south-east coast of the USA.[14] Its name is something of a misnomer, given that its orange-red and cream stripes are more reminiscent of a tiger than a lion, and it is in reality two distinct species of closely related fish. These venomous invaders have no natural predators outside of their native range of the Indo-Pacific, and their numbers have proliferated. In the Gulf of Mexico and Caribbean, the lionfish consume many small fish relied upon as food for local predatory fish like groupers and snappers. Lionfish also feed on fish that are herbivores, those which clean the coral of its surface algae and thus help maintain the health of the reef. Some marine environments have become pervasively infested by the invaders, local animals and plants being pushed to the edge of existence.

The infestation of San Francisco Bay

James Wilson Marshall was unlucky in both love and money. He left his native New Jersey after twice failing as a suitor. As a farmer in Missouri in 1844 he became ill and was advised by his doctor to seek out a better climate. His second farming venture in California also ended in disaster, when having returned from the army during service in the Mexican–American war he found all of his cattle gone. He then entered into a partnership to build a sawmill, constructing one near the village of Collumah on the south fork of the American River, about 180 kilometres to the north-east of San Francisco. On 24 January 1848, he was examining the tail-race from the mill when he noticed some shiny metallic specks, which on further inspection turned out to be gold. News of the 'discovery'[15] spread very quickly—Marshall had told his workers at the sawmill of the find—and, unsurprisingly, the lumber venture was soon doomed, as everyone turned their attention to hunting for the Californian El Dorado. Marshall never benefitted from his discovery and was soon forced off his land by the influx of prospectors. His further business ventures, a vineyard and a gold mine, also ended in failure. Dying almost penniless as an eccentric recluse in 1885, Marshall fared better in death, when 9,000 dollars was raised to erect a monument to commemorate his historic find. But even then, he was to be dispossessed in a final ignominious insult. The original piece of gold that he found in the tailrace now resides in the collections of the Bancroft Library at the University of California, Berkeley, where it is named the 'Wimmer Nugget', after Marshall's assistant Peter L. Wimmer.[16]

The discovery of gold in California was to have much wider implications, heralding the arrival of some 300,000 prospectors—half arriving by sea—and producing a rapid decline in the indigenous population from a mixture of disease and cruel competition for land. San Francisco now became a boomtown, growing from a village to having over 30,000 inhabitants by the early 1850s. As the city grew apace, so too did its appetite, and oysters were introduced to the bay from the east coast. These shellfish fisheries eventually failed, but other marine creatures, from the Atlantic and from across the Pacific, found opportunities to colonize the waters of the bay. Slowly but surely, their invasion began to gather force.

One of the invaders to San Francisco Bay is a chimaera. Its elongate body and glistening flesh make it look like a worm, but the naval shipworm—indigenous to the Atlantic Ocean—is anything but. At one end of its body there are a pair of shells, which it uses to form a tight grip on its burrow. These shells are a highly modified form of the typical seashells you can find on a beach, for the shipworm is a mollusc in which the body has become exaggeratedly long relative to the shell. For centuries the naval shipworm had been boring its way into the hulls of wooden ships. It became such a nuisance to the British Royal Navy that, when faced with the prospect of simultaneous war with the American colonies, France, Spain, and the Netherlands at the end of the eighteenth century, the navy ordered all of its ships to be 'copper bottomed'—expensively lined with this metal—to keep the ships afloat and the worms out. Over a century later, when naval shipworms invaded San Francisco Bay in 1913, it was the American Navy that was to be severely tested. The navy had originally chosen the north end of the bay to avoid attack from shipworms. In that part of the bay the water is brackish, being less influenced by the seawater from the south. For about half a century this strategy kept the Pacific shipworm—the native species of that ocean—at arm's length. But not the naval shipworm. When it entered San Francisco Bay it rapidly spread to the north, being a species tolerant of waters of different saltiness. By 1919 this prolific invader was gnawing through the wooden wharves, demolishing a major structure fortnightly, and continuing this level of destruction for two years.[17]

The naval shipworm is not the only foreigner in San Francisco Bay. The bay has become one of the most heavily invaded aquatic ecosystems in the world, and one of the best studied ones. In some parts as much as 97 per cent of the species are invaders, and they may form as much as 99 per cent of the living mass.[18] They include the Amur River clam, whose home is far away on the other side of the Pacific, in the river that marks the boundary between Asiatic Russia and China. Amur River clams were introduced to San Francisco Bay in 1986 and in some places have proliferated to numbers of 10,000 individuals for every square metre, invading the living space of home-grown organisms. These clams are much more than just space invaders. They are greedy feeders, sucking tiny phytoplankton from the water with almost miraculous efficiency, so that other animals have little to feed on. In San Francisco Bay you might also encounter Chinese crabs, Atlantic periwinkles, Manila clams, and Pacific oysters, all brought in through the direct or indirect action of humans. These are the visible signs of an altered ecosystem. But even at the microscopic level, things have changed.

Like a tiny version of the human-eating protagonist in the 1958 sci-fi movie *The Blob*, foraminifera are amoeba-like organisms that feed using a gliding motion that allows them to form small grasping structures called pseudopodia—false limbs—that engulf their prey. They live in lakes and seas, and like amoebae their bodies are composed of just a single cell (though sometimes this can grow to a couple of centimetres across). Foraminifera have gone one better than their amoeba cousins, and rather than exposing their body to attack from predators, they build a tiny shell for protection. Lurking in large numbers at the bottom of San Francisco Bay is a foraminifer that is native to the coastal waters of Japan. Like many invaders it probably made its way across the Pacific in the ballast tanks of ships, to be unceremoniously dumped into its new environment when the ship arrived in port. This minuscule invader from the Japanese seas (too obscure to have a common name) builds its shell from the available sediment at the bottom of the bay, sticking together individual grains of sand with its own mineral glue. In those parts of the Bay where human influence on the bottom sediments has been very strong, it can account for nine out of every 10 of the foraminifer individuals counted. It is a revolution in this microscopic world.

Jekyll and Hyde species

While rivers and bays, and estuaries and seas are invaded, as in San Francisco Bay, so too is the land. In this modern 'Great Global Interchange', almost all areas of the Earth are fair game for invaders. Many introduced species do not cause harm in their new habitats and may co-exist with the natives benignly or even with benefit. Other animals and plants, though, take on a Jekyll and Hyde existence, useful and productive in their home territories, but bringing death and destruction to the places they invade, just as their human carriers did when they arrived in new lands.

In South America lives the Patagonian bumblebee—*Bombus dahlbombii*. It is a beautiful, orange-coloured bee, the queens growing to 4 cm long. Once it was so widespread that local people referred to the bees as flying mice.[19] It has plied its trade in this landscape since the time of the Inca, and long before that, occupying these lands as the single South-American representative of the bumblebee diaspora, and pollinating the local plants. Or so it was until late 1982 when the 'large garden bumblebee'—*Bombus ruderatus*, originally from Europe—was introduced into Chile from New Zealand and quickly spread. Fifteen years later a second bumblebee, the European 'buff-tailed bumblebee', *Bombus terrestris*, was introduced to pollinate tomatoes grown in greenhouses. A year after that, it was trialled in the open as a pollinator of avocados, for which there was a growing and lucrative market in the kitchens of Europe. The buff-tailed bumblebee seized its chance of freedom. Acting as a villain in its adopted landscape, it spread extremely quickly, reaching as far south as Tierra del Fuego, and extending from the Atlantic to Pacific coast of South America.[20] This bee is projected to invade most of South America.

The deliberate introduction of the buff-tailed bumblebee into Chile has been disastrous for the native bees. Wherever it has invaded, the native bumblebee has declined and is now locally extinct. Buff-tailed bumblebees carry diseases which have infected both the native bees and the introduced large garden bumblebees. They also steal the nectar of flowers that are normally pollinated by hummingbirds.

On the other side of the world, another Jekyll and Hyde species wreaks havoc. The golden apple snail, up to about 6 cm in height and with a thick brown shell, is native to Argentina and Uruguay, where it is considered harmless. The eggs of this snail are bright pink, and it lays these in clusters that resemble a bunch of grapes. A single clutch may contain as many as 1,000 eggs, and the snails are very successful at colonizing new lands. Golden apple snails have spread far beyond their homeland, and can be found in Hawaii, North America, and parts of Europe and the Middle East. But in the wetlands of East Asia they have become a major pest.

These snails were first introduced to East Asia through Taiwan in the late 1970s and are widespread from China to the

Philippines. They have rampaged through the islands of Indonesia too, crossing the Wallace Line with impunity, and reaching Papua New Guinea. They feed on aquatic plants like rice, and so readily make their home in paddy fields. Many initial introductions of this snail were deliberate, as a potential food source for humans, although this role was only fulfilled in southern China, where the snails are eaten raw as a delicacy.[21]

Golden apple snails are typical of many invasive species in that they are resistant to human pollution, adapt well to modified ecologies—like rice fields—and can even survive in waters that suffer bouts of low oxygen. It is the young seedling rice plants they devour voraciously, as they slither through every square metre of rice field. In their homelands of Uruguay and Argentina these snails are culled by natural predators like kites, which accumulate piles of spent shells beneath their perches. Beyond their native ranges the snails have become food for rodents, birds, crabs, fish, and even leeches. They may be cannibalistic, too, the adult snails eating the juveniles. Even with this range of new predators, attempts to remove golden apple snail infestations have proved ineffective. The impact on people, through loss of rice fields, the parasites they carry (like the rat lungworm which can cause a fatal form of meningitis), and reduced biodiversity, as the snails decimate aquatic vegetation and clog up lakes, has been profound.

Animals are not the only Jekyll and Hyde species. Across the world there are many plants with names like 'Devil Weed' and 'famine weed' that reflect their impact as invaders. On the Serengeti-Mara of East Africa one such plant is *Parthenium hysterophorus*, a native of the Americas. This 'famine weed' uses our communication systems against us, spreading through the disturbed land beside roads and highways. Famine weed contains a chemical that causes skin diseases in cattle and breathing difficulties in humans. It is, though, only one of more than 200 introduced plant species on the plains of East Africa.[22]

Another invader has used human communication systems as never before to spread death, damage, and disruption across the world—specifically to human ecologies—in a matter of months. This is the virus SARS-CoV-2, which causes COVID-19. Where the virus arose remains a matter of conjecture, with bats cited as the possible origin, whilst chickens and pangolins may be the intermediaries, the passage to humans perhaps occurring in the Chinese city of Wuhan, late in 2019.

Viewed through an electron microscope, SARS-CoV-2 looks like a spiky ball. This tiny package of calamity is remarkably tenacious, able to survive outside of a host for several days, and on many different surfaces. Spreading quietly through the human population, and in many cases causing no symptoms at all, within a matter of weeks this virus had spread through the population of Hubei Province in China, and then via aeroplanes had become an alien invader across the world, leaving many human tragedies in its wake.

The tiny and invisible virus unleashed a rapid change in human behaviour. People stopped travelling to work, meeting in bars, talking in parks, flying in aeroplanes, and watching football matches. For many months, there were no football matches to watch. As the global economy slowed, oil prices plummeted, and less carbon was, briefly, released into the atmosphere. Grass verges began to grow as lawnmowers fell silent. Within these tiny meadows, insects flourished, and birds and bats that fed upon them grew a little fatter. As the world of humans slowed—albeit briefly—nature began to bounce back. Perhaps this is a brief glimpse of the future—although, as we write, with recovery from the pandemic still uncertain, it is too early to assess long-term consequences.

Long reach of the Homogenocene

In 1999, the entomologist Michael Samways introduced the term 'Homogenocene'.[23] This was a year before Paul Crutzen's on-the-spot improvisation of the word 'Anthropocene', for a geological epoch of the present that is indelibly marked by human actions.[24] The Homogenocene was driven by a similar intuition, that very large changes were taking place around us that were not only impacting our planet at present, but would long reverberate, ultimately to have profound long-term consequences. The word-ending '-cene' implies a geological timescale (think of 'Pleistocene', for instance), with time units typically measured in millions of years. Samways had this in mind when noting the global, human-driven, merry-go-round of species. Occurring at a speed and scale greater than any other biological mixing event in our planet's 4.5-billion-year history, this was not only eroding or 'homogenizing' the kind of distinctive regional biological communities that Alfred Russel Wallace had described; it would affect biology on Earth into the far future too. Indeed, its effects, in changing the course of biological history, will likely persist until the end of life on Earth, perhaps a billion years hence.

For, the Homogenocene world, uniquely in our planet's 4.5-billion-year history, has seen a thorough reshuffling of life on Earth, through the transplanting of species between *every* continent and *every* ocean. It is now a world in which Wallace Lines, and Philip Lutley Sclater's 'areas of creation', have been blurred, and in many places scrambled. These ecosystems will now need tens to hundreds of millions of years to slowly establish their own characteristic natures and identities (whatever those might turn out to be), just as happened with the floras and faunas of Australia and South America after the Gondwana supercontinent began to break up about 175 million years ago. Just as in those biological assemblages, the fingerprints of the Homogenocene will remain in the anatomical structures of the animals and plants of the far future.

Value of the outcasts

There is a more contemporary significance to the Homogenocene: it might, quite literally become a life insurance policy, in the most profound sense, on a changing Earth. Nowadays, the introduction of new species to the landscape is not treated with the blithe abandon that was current in the days of the great Victorian-era explorer botanists and zoologists, as they brought their living trophies back from the distant regions of the world. Today, the damage done to native ecologies is all too clear, so great efforts are now expended to keep alien species from arriving—and in trying to extirpate invasive forms that have arrived. There have been costly programmes (some eventually successful, at least for a while) to remove rats and goats from ocean islands, while the costly wars against ecological invaders such as the Japanese knotweed, the zebra mussel, and the rabbit continue unabated.

And yet, might this extraordinary effort, and the wish to preserve and nurture what is left of native species, be misguided?

Some conservationists note that many of the imported species increase local biodiversity, and not all of them seriously endanger local species. This in itself, they say, is a good thing. A more diverse biological community, regardless of what the components are and where they came from, is beneficial, as it increases the resilience of ecological communities. And resilience will be needed, because the Earth System that supports all life is now a moving target, as the Earth's climate begins to heat up beyond the limits that it has known over the past several million years.

In this scenario (again, still not quite inevitable, but now far more likely than not), traditional conservation cannot work, as 'native' species find the climate around them changing beyond their tolerance limits. In such circumstances, animal and plant communities migrate to stay within their preferred climatic belts—but this is not always possible, as communities are pushed to the equivalent of a local cliff edge (say, to the edge of a continent or the top of a mountain). In any case, it is made hugely difficult because the remnants of 'natural' biological landscapes have been fragmented into pockets within the growing envelope of farmlands and urban areas.

The Australian husband-and-wife team of ecologists Maggie and David Watson[25] make the point that it is the introduced species—currently maligned, chased, harried—that have the best track record of surviving new and difficult conditions. In picking a path through shifting climatic belts to come, they may be best placed to form the basis of a functional biology on a world (likely a much-degraded world biologically) after humans have become extinct. Tough, adaptable, and mobile, the likes of the cat, rabbit, and rat, and golden apple snail, prickly pear, and Amur River clam, may be the best bet for the post-human renaissance.

To the Watsons, it is a kind of solace, and one that, in a small and local way, they try to put into practice. They have 'adopted' their own local favourites among the outcasts, in studying mistletoes, and gulls, and tern—organisms normally thought of as pests that need to be controlled—and in educating the next generation of conservationists about the dilemmas they will face and the decisions they will have to make. We cannot predict the world to come, but we can try to tip the balance in favour of life.

4

THE LOST WORLDS
OF THE GIANTS

Walk into a natural history museum anywhere in the world and the visions that greet you are nearly always of giants. In Washington DC, *Tyrannosaurus rex* is locked in a death struggle with a *Triceratops*. In London, the skeleton of the largest animal that has ever lived—the blue whale—seems to glide towards you, suspended from the roof, while the Dinosaur Park at Beijing's Natural History Museum has no fewer than 23 reconstructed behemoths roaring, waving armoured tails, and (for extra realism) panting and blinking.

It seems the human psyche is drawn towards the awesome, the spectacular, the terrifying, even the monstrous. This attraction, seemingly at odds with our instinct towards self-preservation, was present long before the age of modern, well-stocked museums. In the cabinets of curiosities that could be found in the mansions of the aristocracy of the seventeenth and eighteenth centuries, fossil mammoth teeth and tusks were 'must-have' items. A particularly celebrated, indeed notorious, prize was a magnificent, wickedly toothed mosasaur skull discovered in Maastricht in 1770, that became a prime target for Napoleon's forces (on Bonaparte's direct orders) when they laid siege to that city in 1794. With the help of 600 bottles of wine as a bribe—as some stories say—it was duly carted off as war booty to the Natural History Museum of Paris, where it still remains (Maastricht would like it back but has not succeeded in extricating it yet). Even further back in human history, in Greek and Roman times, the bones of giant extinct mammals were avidly sought after, and sometimes sparked a 'bone rush', as these were seen, then, as precious relics of the ogres and heroes of legend.[1]

Within the labyrinthine corridors and spaces that generally lie behind the grand museum galleries, the rest of the more humdrum petrified life is stored. These storehouses are in their own way just as spectacular, just as breathtaking as the dinosaurs on display for the public. One such storehouse is the British Geological Survey's museum at its main headquarters, just to the south of Nottingham, where fragments of life that populated the seas and land millions of years ago fill its many storage cupboards. This is the oldest geological survey in the world, and since 1835 its scientists have been cataloguing the nation's rocks, and the riches that lie within them, including the wealth of fossils they contain. There are now some 3 million fossils in its collections, in racks of purpose-built wooden trays four levels high. The wooden trays contain smaller cardboard boxes in which the fossil-bearing slabs are carefully stacked and numbered, each slab separated by a protective strip of paper from its neighbour. Some palaeontologists, conscious of space, packed these slabs in so tightly that an incautious attempt to pull one slab out can cause a whole boxful to spring out into the air, in a small palaeontological explosion. It is a goldmine of extinct life that seems endless (though it is dwarfed in turn by the 40 million fossils of Washington DC's Natural History Museum).

In some of the larger storage cupboards are the bones of hyena, rhinoceros, elephant, and hippopotamus that walked the landscape of Britain just a short while ago geologically, most being from the last-before-present warm climate interval, about 120,000 years ago, and from the 100,000-year interval after that when ice or permafrost mostly covered the land. Some of these bones—hyena bones, notably—come from the caves of Creswell Crags which lie in a narrow gorge that straddles the county boundary of Derbyshire and Nottinghamshire. Here, on the northern habitable edge of Ice Age Europe, people and animals vied for the life-protecting cave space, leaving behind their remains over a period of nearly 50,000 years. Artefacts from these caves include stone tools made from flint, engravings on the walls, bone implements including sewing needles, and the bones of animals that once lived here, including those of humans. On the northern side of the gorge, on its sunny south-facing slope, is the labyrinthine Robin Hood's Cave, where that legendary and probably mythical outlaw is said to have hidden away from the Sheriff of Nottingham's men. Immediately opposite, on the southern side of the gorge, is the long narrow cave called 'Church Hole'. It shows the faintest traces of bas relief cave markings, perhaps once painted with ochre, but now often covered by a thin layer of limestone that has grown over the cave's surface from thousands of years of dripping water. In the dim light of the cave, deer, bison, auroch, and other animals emerge as ghostly outlines on the walls. At the end of the last ice age many giant animals, like the woolly mammoth and woolly rhinoceros, still occupied the British landscape—but they soon disappeared. It was a pattern that was being played out across the world.

Demise of the giants

The caves of Cresswell Crags are among the latest part of a long historical record left as fragments of bone and scattered stone-tool artefacts across the Old World. For many millennia, ancient humans existed as a small and ephemeral part of a landscape which they shared with many other animals and plants. Even so, as early as 2 million years ago they had developed the ability to kill and process large animals like antelopes and horses.[2] Archaic humans had also developed a command of fire, perhaps as early as 1.5 million years ago,[3] and over time they gradually evolved stone tool kits of increasing complexity and utility. The first signs that our ancestors, in the form of *Homo erectus*, might have begun to change the animal landscape came in the form of modest extinctions of sabre-tooth cats and some elephant species in Africa, that cradle of humanity, about a million years ago, while these animals continued to thrive elsewhere.

Early human species travelled: one of their number, known to us as *Homo antecessor*, had reached Britain some 900,000 years ago, leaving implements and footprints by the cliffs of what is now the village of Happisburgh in Norfolk. Members of our own species, *Homo sapiens*, were living in North Africa by 300,000 years ago, but still with mostly negligible effects on other species. That pattern began to change from about 70,000 years ago, as cognitively modern humans left their mark on Africa, Europe, and Asia, and then moved on into new lands.

Australia was one of these new lands, where no humans had trod prior to our arrival 65,000 years ago. People crossed from South East Asia during the last Ice Age when sea level was lower, and more land was exposed. This journey must nevertheless have required remarkable navigation skills as they travelled, perhaps via the Timor Sea, or via island-hopping from Sulawesi to Papua New Guinea. At its narrowest stretch, these early mariners must have traversed some 90 kilometres of sea, with any hint of land being far beyond their gaze. What kinds of boats these people used, or how they made the crossing, is unknown. The oldest preserved boats are much younger than this, dugout canoes dating to about 8,000 years ago. Today, it is very hard to imagine anyone paddling across a large expanse of open water, such as the Timor Sea, in such a crude boat. By whatever means these settlers arrived, over succeeding millennia they dispersed across Australia.

The biological landscape that these ancient humans explored, and proceeded to transform, was remarkable. Even the remnants that had survived by the time the first European explorers arrived were enough to provoke amazement, with bizarre creatures, to their eyes, such as the kangaroo, koala bear, and potoroo. Charles Darwin, who set foot in Australia as a 26-year-old in 1836 during the return leg voyage of the *Beagle*, was deeply struck by the sight of platypuses playing in a river on the slope of the Blue Mountains, by Wallerawang.[4] Here were creatures that were occupying the niche of a European water rat, with similar adaptations— but, as mammals that possessed duck-like bills and laid eggs, were almost absurdly different biologically. If species had been fashioned by any sensible Creator, why would not such a deity have simply put the water rat in Australia as well as Europe? That would seem to be much more straightforward than designing a whole new and very different organism. Darwin noted that 'an unbeliever in everything beyond his own reason' might be prompted to cry out, 'Surely, two distinct Creators must have been at work!' It has been called Darwin's 'platypus moment'[5]: a significant step that set him on track to questioning divine creation, and towards building his theory of 'descent with modification' by natural selection.

What might Darwin and his colleagues have thought if they had alighted on to the Australia of 65,000 years ago? They would have encountered beasts that were not only bizarre, but impressively bulky too. There were such as the *Diprotodon*, the 'giant wombat'—a marsupial 2 metres tall at the shoulder and weighing nearly 3 tonnes; *Procoptodon*, a giant kangaroo up to 3 metres high; *Propleopus*, a 70 kg carnivorous or carrion-feeding kangaroo; and *Zaglossus*, an echidna the size of a sheep.

By 40,000 years ago, human populations had reached Western Australia, and as far south-east as New South Wales. The giant kangaroos and wombats were by then extinct, likely killed by a combination of hunting and vegetation burning. These skilled hunters left clues as to their tastes—for birds' eggs, for instance. Fragments of eggshells attributed to a bird that looked like a large turkey belonging to a group called the 'megapodes'—'birds with big feet'[6]—commonly show signs of burning and cooking in a fire. This kind of pattern, of a spread of humans and demise of giant animals, was to intensify around the world in the coming millennia.

The Americas lie even more geographically remote from the origins of modern humans than Australia, but at the time of the last Ice Age there was a direct chain of geographical connection through north-east Asia and onwards via the icy wastes of the Beringia land bridge that joined easternmost Siberia to Alaska. This land bridge gradually disappeared as meltwater from receding icesheets forced sea level to rise at the end of the last Ice Age. People had probably been living in this landscape for 5,000 years when the rising waters dispersed some back to Asia, while others made the journey into the Americas. Before humans arrived there, the biological landscape of North America was very different from what it is now.

Here lived true giants, aptly named 'megafauna'. Along with the more familiar Ice Age animals, such as mastodons and mammoths, there were giant beavers 2 metres long with formidable 15 cm-long incisors, and *Camelops*, a North American relative of the camel that stood 2 metres tall at the shoulder, while the formidable American lion was up to 2.5 metres long. Outgunning all of these for sheer size was the giant short-faced bear, standing 1.8 metres high at the shoulder, and rearing up to be 3 metres tall when standing on its hind legs. Likely a voracious predator, it must have been a fearsome sight even for a hardy human hunter from Beringia. Some of these animals, like the giant beaver, appear to have been in decline before the arrival of humans, perhaps due to climate change at the end of the last Ice Age. But others show clear evidence that the human invaders were hunting them. Amongst these, one group of megafauna had grown to colossal proportions: the ground sloths.

Ground sloths first evolved more than 30 million years ago on the then-isolated continent of South America, where some of their living relatives, tree sloths and anteaters, still live. As the two Americas became joined through the formation of the Isthmus of Panama about 3 million years ago, ground sloths migrated into North America, eventually reaching as far north as Alaska. They also reached into the islands of the Caribbean. Some of these ancient sloths reached elephantine proportions. *Eremotherium*, which was present in North and South America, could grow to 6 metres long and weigh 3 tonnes.

Humans were clearly hunting and butchering these gigantic animals. Near the surface of the playa lake called Alkali Flat in the White Sands National Monument of New Mexico, there are superbly preserved fossilized footprints that are at least 10,000 years old.[7] These show close interaction between sloths, sometimes apparently flailing around for protection, and human pursuers. So close is the connection that the human footprints sometimes stand within those of the sloths. Whether people really were actively hunting them is difficult to discern, because there are no preserved carcasses at the site, and no evidence for butchery. Elsewhere, in South America, there is direct evidence of sloth kills.

At Campo Laborde, in the Pampas of Argentina, is a 12,000-year-old record of the largest of all ground sloths being butchered, a *Megatherium*—the name literally means 'great beast'—which reached 4 tonnes in weight. Many other bones of mammals are found at the site, but only *Megatherium* and the Patagonian hare—a species still living today on the Pampas—show signs of being butchered. There are cut marks on a sloth rib bone that indicate the removal of its flesh, while those on the leg bone of the hare suggest skinning.[8] Although such finds of butchery are rare in the Americas, the rapid disappearance of ground sloths within about 2,000 years of the arrival of humans suggests that hunting drove their demise. Another of these rare sites is La Moderna, also on the Pampas, where a large glyptodont called *Doedicurus* was butchered on the margins of a swamp some 7,500 years ago. This was a heavily armoured animal a little like its modern mammal relative the armadillo, but one that possessed a long tail that bore a spiky club at its termination, like that of the dinosaur *Ankylosaurus*. This club was probably of no use in defending it from humans, as the animal had poor rear vision (more likely, it was used in conflicts with other glyptodonts). It and many megafauna disappeared from American landscapes. The last few ground sloths clung on in the islands of the Caribbean until about 4,000 years ago.

Our fascination with these rare, gigantic animals is not entirely frivolous, nor just a way for museums to boost their incomes. The sheer scale that these leviathans could reach tells us something of the limits of biology, in different times of the Earth's past and in different conditions, on land and in the sea, in good times and bad. These giants fuelled their bulk by consuming the tissues of very many plants, or of many other animals. The apex predators—and the apex herbivores that on land could often reach even bigger proportions—were entirely dependent upon the overall health of the meshwork of life, among which they formed such spectacular outliers. These giants also exerted a control on that landscape which is hard for us to imagine in these days of impoverished wildlife. But let us take stock of the scale of the change that has taken place on Earth, over those thousands of years.

The scale of loss

To assess the scale of loss, we first need to define some categories. What are the megafauna? One simple and commonly used category is just to lump together all moderately to large-sized animals, and have a modest cut-off point, at 100 pounds weight (45.3 kgs). But there is an ecologically based classification that more clearly separates the giants from the rest. Megaherbivores, here, are understood as plant-eaters weighing more than 1,000 kg, while megacarnivores make the grade at 100 kg.[9] Why the order-of-magnitude difference? Well, one good defence against predation is simply to grow too big to sensibly attack, and this has been a constant strategy for herbivores since animal life developed on land.

Once having reached that size, an animal becomes more than an impressive addition to the scenery: it *shapes* that scenery. The megaherbivores change the balance between grassland and forest by prodigious eating, by trampling, by uprooting trees on an enormous scale. Within a forest, they can change tree community structure by more easily destroying saplings, the surviving trees then forming extensive stands of mature forest. The enormous volumes of vegetation eaten and then excreted fertilize the soil—especially as the herbivore guts help break down the often tough and resistant plant matter. Herbivore guts, too, are a good vehicle for dispersing plant seeds, and many fruiting plants have co-evolved with herbivores over geological time, becoming as dependent on them as many flowers are on pollinating bees. These giant herbivores, especially when young or sick, are not altogether immune from the megacarnivores, of course, as these in turn shape the biology around them through the 'landscape of fear' they generate.[10]

That—with only geologically brief pauses (such as just after the dinosaur-killing impact of the asteroid that brought the Mesozoic Era to an end)—has been the state of the terrestrial landscape for hundreds of millions of years. That kind of long-term global functioning of the terrestrial biosphere has, in the geological blink of an eye, come to a halt, as the landscape-shapers have been decimated. Before the main phase of human-driven megafaunal extinctions began, some 50,000 years ago, there were some 50 megaherbivore species in the world. Now just nine remain, all in Africa and Asia, as the still-existing (if severely depleted) elephant, hippopotamus, and rhinoceros species. The Americas have fared much worse, losing all of their 27 megaherbivore species. The 10 largest species of that original 50 have all become extinct, so that even the remaining megaherbivores are modestly sized, as giants go. The consequences of this loss are now difficult for us to envisage, or understand at a visceral level, given the little that is left of this kind of world that we can directly experience. But, these consequences may run deep.

One consequence—particularly noted in North America, where the megafaunal crash was acute, was a conversion from a patchwork of forest and park-like grassland, sustained by all that trampling and uprooting, to more continuous forest cover. And that in turn commonly seemed to have led to more widespread and frequent fires, with the loss of natural firebreaks as the giant animals declined, to produce a landscape tilted towards a 'black world' (frequently fire-ravaged) scenario, and away from 'green world' (forested) or 'brown world' (herbivore-maintained grassland) conditions. Even within a forest, the megaherbivores are thought to help encourage the growth of old, mature trees, grown too large to easily uproot, by thinning out the more vulnerable samplings. This kind of landscape change wasn't just restricted to the tropical and temperate parts of the Earth. At high latitudes, in cold northern Europe and Asia, a constant feature of warm intervals in Ice Age times were 'mammoth steppes' with dry fertile soils that supported a rich animal fauna and a diversity of plants. When the mammoths were hunted to extinction, these conditions of plenty disappeared, to be replaced by waterlogged, much more biologically impoverished tundra-like soils. The key to the mammoth steppes seems to have been the abundant megafauna that, by constant grazing, encouraged the growth of deep-rooted grasses that helped cycle both nutrients and water.[11] With this ecological function thousands of years ago, this 'natural' (to living human memory) landscape of Siberia and kindred regions already represents a heavily modified—and poorer—kind of world.

These wide continental landscapes, thus, have long (on human timescales, that is) been transformed. But some parts of the world long remained essentially pristine and untouched, protected by their geographical isolation. Islands in the middle of great oceans have formed little isolated worlds of their own since the early days of our planet, each one a laboratory for biological evolution for the few castaway species that landed there. But even those refuges for animals and plants large and small, could not escape the constantly spreading web of human impact. In such places, at least, there is recorded human witness for the transformation.

Fragile islands

There are many places on Earth, on the land and in the sea, where animal and plant species have already made their last stands. On others, they are now in the midst of seemingly terminal struggles.

The island of Rodrigues in the Indian Ocean is one such place. It is terribly isolated, lying about 560 kilometres (350 miles) to the east of Madagascar, its total surface area only about 108 square kilometres (40 square miles). Still, this small island has a wide range of terrains rising to nearly 400 metres above the sea on Mount Limon, and before humans arrived, its heavily forested landscape supported a diverse island flora and fauna. Here, until the eighteenth century, lived the Rodrigues Solitaire, a less famous cousin of the dodo. The bird approached the size of a swan, though with a pigeon-like body with long legs and neck, hooked beak, and tiny wings. No longer useful for flight, the wings were used in combat, and also for communication, perhaps to police their territories; the sound from their flapping wings was said to carry for some 200 metres. The first person to describe the solitaire was the naturalist François Leguat, a French Huguenot escaping persecution in France, who was marooned on the island for two years in 1691. He seems to have been quite taken with the birds' appearance, which he described as moving with grace and stateliness. This, though, did not save the birds.

The Rodrigues Solitaire, a solitary bird on a solitary island, had nowhere to escape to when the invaders came. Its forest habitats were destroyed, the birds were hunted for food, and their eggs were probably consumed by pigs and cats. By the middle of the eighteenth century, a few decades after the first sighting of the birds, they were gone. The Rodrigues Solitaire was not the only casualty.

Before humans arrived on Rodrigues there may have been tens or even hundreds of thousands of giant tortoises. These were hunted to near extinction during the mid-eighteenth century for food and fuel. There are still giant tortoises on Rodrigues now, but these are introduced Aldabra tortoises, another giant species of the Indian Ocean, and one that is important for engineering ecologies for other organisms to live in, called tortoise turf. The indigenous tortoises of Rodrigues may have filled a similar function, dispersing seeds and making clearings in which diverse plants could thrive. There were two native species, the giant 'saddlebacks', which lived on the taller vegetation, and the smaller 'domed tortoise', which was adapted for feeding on undergrowth. By the beginning of the nineteenth century only a few surviving tortoises could be found, but too few for a breeding population.

Other Rodrigues animals suffered the same fate, as the island was cleared for agriculture by burning its vegetation, and by the combined onslaught of cats, rats, and pigs. The vanished animal species include the Rodrigues starling, an elegant bird with white plumage and a yellow beak, which seems to have clung on in some tiny islets that surround the main island, probably until rats swam across from the mainland and set about eating its eggs. Gone too is the beautiful green-backed 'Rodrigues day gecko', which lived amongst the trees where it fed on insects and nectar. Five of these animals were pickled in museum collections for posterity, an ignominious end to a lizard that reached a quarter of a metre long. What happened on Rodrigues is typical of the islands of the Indian Ocean, and those of the Pacific and Atlantic too, many of which have lost thousands of species.

Islands beneath the sea

The Earth's wide landscapes have become biologically diminished since human written records began—and long before. Even what were once considered wild, pristine forests and savannahs were fundamentally altered as a new top predator, *Homo sapiens*, reconstructed the apex of the terrestrial ecosystem, and set in motion a cascade of ecological consequences. This is a large part of what makes the current interglacial phase quite distinct from the fifty-odd preceding interglacial phases of the 2.6 million year-long Quaternary Ice Age, and helps justify setting it apart formally as its own epoch, the Holocene, on the Geological Time Scale.

In the oceans, however, things were different. For a long time, these deep, dangerous, and seemingly endless expanses of water excluded any kind of fundamental human influence. Humans did cross them, with courage and skill, to reach new lands, and they did fish, though mainly in shallow and coastal waters. But, for most of recorded history, they barely scratched the surface of the main bulk of the oceans, and the richness of life they contained.

A few centuries ago, serious onslaughts on this realm began, as fishing boats were developed that could make longer voyages, and tackle bigger prey, up to the oceanic megafauna of whales, dolphins, and sharks. The decimation of these charismatic beasts, and of the smaller fry lower in the food chain, peaking with the industrialized fishing fleets of the nineteenth and twentieth centuries, is—unlike the ancient disappearance of the land giants—documented with statistics and eyewitness accounts, and vividly chronicled.[12]

In this short time, the deep oceans have been transformed from seemingly endless and inexhaustible resources to being, now, in many places essentially fished out. The reach of technologized humanity now extends through the increasingly impoverished waters, and is set to extend to another set of islands, those that lie beneath the sea.

Out of sight, far beneath the surface waves of the ocean, these other 'islands' have their own specific ecosystems, often harbouring unique forms of life. These are the islands made by ancient and long-dead volcanoes that form underwater seamounts. They too are now threatened, by deep-sea fishing and mining. There may be as many as 100,000 of these submerged islands with their biodiverse communities, that together cover a huge area of the oceans.[13] Although their physical distribution has been generally mapped by geophysical surveys, only a few hundred of them have been studied biologically in any detail.[14] Many of these seamounts lie beyond the jurisdictions of individual countries and thus are threatened by overfishing and exploitation in the high seas.

One such area of seamounts is the Walters Shoals that lie about 850 kilometres to the south of Madagascar in the southern Indian Ocean at latitude 33° 12' 0" S, longitude 43° 54' 0" E. Geographically these seamounts are part of the Madagascar Ridge, that lies to the north of the south-west Indian Ocean ridge, a 6,000 kilometre (3,700 mile)-long gash in the Earth's surface where new ocean crust is being formed by volcanic activity. The Walters Shoals are mountains that rise more than 4,000 metres from the seabed and are at some points just 15 to 18 metres below the sea surface. Farther down, at 500 metres depth, the flat-topped surface of the shoals covers an area of about 400 square kilometres, or four times the area of Rodrigues. The Walters Shoals has a unique 'island' fauna of sea lilies, sponges, crabs, shrimps, corals, lobsters, and fish. And because the upper reaches of the shoals are in the zone that is penetrated by light, there are pink reefs of coralline algae—communities of small unicellular organisms that are capable of making large limestone structures.[15] Many species of these have so far been identified on the shoals.[16] The seamounts also provide a habitat for whales, whilst seabirds gather here to forage. In a story that shows parallels with the destruction of Rodrigues' biodiversity, the Walters Shoals have subsequently been trawled and fished. Not discovered until 1962, the shoals once supported a large population of Galapagos sharks, but these were rapidly fished out.

The Walters Shoals are just one part of a much bigger problem facing ocean islands below the sea. They are now seen not only as a source of fish, but also of precious metals like cobalt and tellurium. Many deep-sea mining companies are keen to prospect for these metals, which form the basic components of batteries for electric cars and solar panels, a supposed solution to our over-consumption of fossil fuels. But the collection of these metals from the seabed involves the use of giant suction devices, or continuous chains of buckets, that rip through the delicate ecosystems. It is part of the dilemma we face. Does our generation continue to consume the Earth's resources in this kind of way, and degrade the rest of nature, making a mass extinction inevitable? In such a scenario, the oceans and land will have an impoverished assemblage of organisms for (at least) hundreds of thousands of years into the future. Or are there ways to live *with* nature and preserve its biodiversity? The nature of past mass extinctions, and the millions of years it took for life on Earth to recover after each of these events, shows what is at stake.

The day of the dead

On 2 November each year Mexicans celebrate the 'Dia de Muertos'. The festival is an old one with its origins in the Aztec goddess Mictecacihuatl, the keeper of bones in the underworld of Mictlan. On this night in November a blurring occurs between this world and that of the dead, and the spirits of the deceased are able to join the world of the living again. For most of Mexican history this festival was mainly celebrated in the south of the country, while today in Mexico it has become a national holiday. Its fame has been enhanced in the most contemporary of ways. The 2015 James Bond film *Spectre* featured an (entirely mythical) *Dia de Muertos* procession in Mexico City. This magnified the festival's notoriety worldwide, and the government saw it as an opportunity to highlight indigenous traditions. Since then, an annual procession has taken place for real through the streets of the capital.

The *Dia de Muertos*, though—even without its recent association with a licence to kill—is not a celebration of death, but one of life, in invoking the spirits of the past to join in with the lives of their living relatives. For palaeontologists, all animals, plants, fungi, and microbes that have ever lived also fall within the scope of such a view of the day of the dead, as ever more ingenious attempts are made to try to bring their fossilized remains back to life within our imaginations. Some of these organisms, like the giant *Tyrannosaurus rex*, are remembered from their petrified skeletons. Others may leave impressions, footprints, burrows, or trails, while yet others have been transmuted into coal, oil, and gas, and now join us in our lives in the most literal of ways. Very many—indeed most—of these organisms have not yet been detected, especially if they have left no physical trace of their own bodies. Yet, scientists still try to resurrect some echo of their activity: a virus might leave no direct trace of itself in strata, for instance, but its activities might be inferred from patterns in the fossilized remains of the animals it infected.

At the heart of these painstaking reconstructions is the desire to understand the grand patterns of life and death on Earth. Life and death, of course, can be understood in different ways. We usually think of it in relation to individual lives. No organism has lived forever on planet Earth, though some can have a good innings. Creme Puff, the Texan cat, lived for 38 years, for instance.[17] Her owner put this down to a healthy lifestyle and a good diet of broccoli, turkey, bacon, eggs, and coffee with cream, with a snifter of red wine. Thirty-eight years is a long time for a cat, though far outdone by Ming the 507 years old Atlantic mollusc, unceremoniously frozen to death (by humans) in 2006. Poor Ming,[18] in turn, continues to be far outlived by Methuselah, a 4,852-year-old (as we write) bristlecone pine growing in the White Mountains of California. It first sprouted at about the time that cities emerged in Mesopotamia, and before the Pyramids were constructed. However, it too is a mere youngster compared to the tiny bacteria occupying the ancient ice of Antarctica that may have been alive—if in a state of near complete suspended animation, since the Pleistocene.[19]

The longevity of some individuals, at least, can approach geological timescales. But what about the longevity of individual species, like rabbits or dogs? Here we are in deep waters from the beginning, for a rule-of-thumb test for a modern species—organisms that can interbreed to produce fertile offspring—cannot be recognized for fossils. So, palaeontologists use 'morphospecies', defined by distinct and consistent shapes of fossilizable parts like bones or shells—a definition which usually

requires a lot of careful observation and measuring in practice. With that proviso, what might the geological record of a
Nile crocodile be, for example? Its fossil record extends back to
the Pliocene Epoch about 3.5 million years ago,[20] and so one
might regard it as a long-lived species. Other species had briefer
sojourns on Earth: many species of the beautifully coiled ammonites that lived in the seas of Jurassic and Cretaceous times
lasted for only a few hundreds of thousands (and some seemingly just a few tens of thousands) of years: their early demise
in those times now makes them hugely useful to geologists, as
precise time-markers for the strata that they are found in. At the
opposite end of the spectrum of both size and time, the tiny
single-celled foraminiferan *Trilobatus sacculifer* seems to have originated in the Miocene Epoch, some 20 million years ago.[21] It has
left a worldwide fossil record, because it lives in the ocean waters and its tiny calcareous skeletons sink to the seabed on death.
The Nile crocodile and this foraminiferan continue to thrive, but
one day they too will join the long-vanished ammonites. Species
thrive because they have evolved to be well adapted to their environment. If this is large, like the global tropical oceans, they
can thrive for a very long time. But if their habitat is small, like
a tiny island, then their chances of becoming obsolete are very
much greater. Other species thrive because they are generalists,
like cockroaches, and so can adapt to many different landscapes,
including those strongly modified by humans.

Many palaeontologists have tried to come up with a figure for
the average durations of different species, for groups like mammals or molluscs. Species longevity seems to vary considerably
depending on the group, and the quality of its fossil record. But a
figure of somewhere between 0.5 to 5 million years is thought to
be the typical duration of most species. These calculations allow

us to estimate the rates of extinction over time, called the 'background extinction'. By such estimates it would be expected that nine vertebrate species would have gone extinct since 1900.[22] But background extinction rate is not a constant and is influenced by factors such as climate change, super-volcanoes, and asteroid strikes. Sometimes these factors can have a colossal influence, and trigger mass extinctions of species.

The killing times

Mictecacihuatl has stacked up the bones of most of the species that have ever lived on Earth, and which are now extinct, and she has assembled these for us in the fossil record. Animals and plants eventually go extinct because other organisms become more competitive in their environments, or because the environment itself changes and they cannot adapt. Species with narrow environmental ranges, like many island species, are more susceptible to extinction, and conversely this is why cockroaches, rats, and jellyfish are good survivors. The rate of extinction of plants and animals is worked out from the average timespans of species, based on observational data, and from a reading of the fossil record. From such information it is possible to make an estimate of background extinction rate, and this suggests we would expect to lose one mammal species every 200 years, and one bird species every 400 years. During periods of mass extinction this changes dramatically, and millions of species may be lost in a short time— some in the space of a single day, as when the Cretaceous world came to an end.

There have been five such episodes of mass extinction in the past 500 million years, and if we reach back a little further to encompass the loss of the strange Ediacaran organisms about 539 million years ago, one might add a sixth. Each time the Earth has lost something of the order of 75 per cent or more of all its species within a geologically brief interval ('geologically brief' meaning anything from several million stressful years to perhaps just a few months at the end of the Cretaceous). These extinctions, and many lesser but still important ones, punctuate the fossil record. Each major extinction pulse has played out over different frames of time and space, and different kill mechanisms, too—albeit with some common factors.

The more ancient major mass extinction events played out almost exclusively in the seas, for the simple reason that there was little then in the way of life on land. The 'pre-Big Five' event at the end of the Precambrian that saw the loss of many Ediacaran organisms may have been the result of evolution. For the first time on Earth, an array of mobile, muscular, skeleton-bearing predators emerged for which the Ediacaran organisms were easy prey. These new mobile animals also burrowed through the microbial mats that many Ediacarans relied on. The Ediacarans may have been the first victims of the arms race between hunter and hunted that goes on to this day.

That first of the 'Big Five' mass extinction events within the Phanerozoic took place at the very end of the Ordovician Period, some 445 million years ago. It was a puzzling and unique event in a number of ways—but there are also ways in which it eerily resembled our own contemporary crisis. It seems to have been an episode of mass death propelled not by mighty volcanic outbursts or catastrophic meteorite impact but rather by an extreme oscillation of climate. Within the space of a million years, an enormous ice sheet grew over much of South America and southern Africa, then conjoined and positioned over the South Pole. It abstracted so much water from the oceans that much of the continental shelves went from being shallow seas to being dry land.

The ice sheet was short-lived, though, and collapsed as a pulse of global warming led to flooding of oxygen-poor water back across the shelf areas. Long interpreted as a kind of double whammy for the biosphere, it has now been suggested that the initial climate shock did most of the damage, with the abrupt reflooding simply complicating the recovery process.[23]

This would make sense. The shallow sunlit waters of the continental shelves were the main cradle for the abundant and diverse life that was developing in the late Ordovician, as the land was still largely barren except for some mosses growing in moist places, while the ocean depths were often starved of oxygen. The sharp reduction in the extent of continental shelves as the ice sheet grew simply left less room for life. In the firing line were most of the coral species then living, and their disappearance led to one of the first 'reef gaps' in Earth's history.

There is a sense also of a loss of the most flamboyant and sophisticated forms of life. This was the heyday of the charismatic, carapace-bearing trilobites. Not all of these succumbed, but among those that did were many of the more elaborately constructed forms. Late in Ordovician times, for instance, there thrived many species which developed an elaborate fringe, marked by complex arrangements of holes, around their heads. These abundant and mysterious animals (palaeontologists still scratch their heads over the purpose of the bizarre fringe) all became extinct as the ice grew. There were also trilobites which had left the sea floor to join the plankton, with light, thin armour and huge eyes that extended all around their heads, so they could see things below them as well as to the sides and above: the ideal animals to survive an extinction that reduced sea floor space, one might have thought. But no, these disappeared too (and never re-evolved), in company with other specialized forms of zooplankton. There was clearly an array of kill mechanisms at work, reverberating through the whole ocean system as continental shelf space was squeezed. The survivors tended to be small and simple forms, the kinds of animals that one can loosely interpret as generalists. Through the bad times, these hung on. Some even thrived, and have been termed 'disaster species' for the effectiveness with which they coped with conditions that killed most other species. Even in the worst of times, there have been winners as well as losers.

Later in Earth history, once life spread in force to the land, mass extinctions had a wider theatre in which to operate, and other sets of factors came into play. Since that time, curiously, there has not been another *major* mass extinction triggered by a glaciation. This might have been because the End-Ordovician glacial pulse was particularly fierce. But it may reflect change to the Earth's biological fabric too. Life on land is not quite so sensitive to sea level change as regards living space (it can migrate downhill as sea level falls, for instance), and it is significantly less sensitive to temperature change, having to cope with daily and seasonal air temperature rises and falls greater than those in the sea. Subsequent onsets of glaciations have seen some groups of animals and plants suffer—but nothing on the scale of the end-Ordovician calamity.

The more recent 'great dyings', which affected both marine and terrestrial life, have had different triggers. By far the most notorious one is that at the end of the Cretaceous Period 66 million years ago, when the dinosaurs and much else perished. This was long among the greatest mysteries in geology, until the discovery by the father-and-son team of Luis and Walter Alvarez of a thin layer of iridium-rich dust at the exact stratal layer where the extinction event is clearly signalled by a drastic change in the kinds of marine microfossils present. The iridium layer was later found in rocks worldwide, and also contains frozen droplets of melted rock and intensely shocked mineral fragments. It most plausibly came from a giant meteorite impact, and a 200 kilometres-wide crater of that age was later found beneath the Gulf of Mexico.[24]

A clear case of biosphere collapse by catastrophic impact, one might think. And yet, there was (and remains in some quarters) considerable resistance to the idea. In some places the pattern of extinction seemed gradual rather than sudden, which might suggest a gradual kill mechanism, such as a powerful series of volcanic eruptions—which did occur around that time, in what is now the Deccan region of India. And, there is strong evidence that such extraordinary volcanism *can* cause mass extinction events, as the gases released poison the air and sea, and change climate. That is now clear for the greatest mass extinction event known, at the end of the Permian Period 251 million years ago, when more than 90 per cent of species died out, and coral reefs were lost from the oceans for several million years, and for another mass extinction event at the end of the Triassic Period, 200 million years ago. So why not the end-Cretaceous event too?

It has taken a great deal of detective work to show that the end-Cretaceous meteorite likely *was* the main culprit. In regions where a more gradual extinction had been surmised (as with dinosaurs on land), this was often shown to be due to the imperfections of the fossil record; where fossils are rare in any case, as with dinosaurs, it is hard to demonstrate that they disappear simultaneously. The pattern of the Deccan volcanism, closely examined, did not fit the pattern of extinction. And, drilling into the crater itself has shown the extent of the mayhem wreaked at the impact site, and made the global repercussions more plausible.[25]

For the past 66 million years there has been no mass extinction. Now, though, the rate of extinction loss is far higher than the background level—and far higher also, with far wider effects, than during the megafaunal extinctions likely caused by our Stone Age ancestors. More than 50 mammal species have been lost since 1900,[26] and these include iconic animals like the Caribbean Monk Seal (last sighted in 1952) and the Yangtze Dolphin (last seen in 2002). If this trend continues, we might expect to see 75 per cent of species lost within a few hundred to a few thousand years,[27] and we would then be in a sixth Phanerozoic mass extinction event. This calamity is not caused by volcanic eruptions, asteroid strikes, or rapid climate change, but by human activities, and these are now so extensive that the whole of the world is threatened, in effect behaving as a single if gigantic island, like a far larger version of Rodrigues, when humans first arrived in the seventeenth century.

We cannot board Noah's Ark and escape this flood of environmental devastation. There are no other nearby islands to which we can flee. Our closest planetary neighbour Venus is a Hadean world with surface temperatures hot enough to melt lead, and an atmosphere so heavy that it would the crush the life out of any Earthly plant or animal. Mars too is an inhospitable place, with no oxygen in its atmosphere and almost no water at the surface to sustain life. As Carl Sagan said, the Earth is where we must make our stand.

The last of the velvet worms?

Despite the periodic setbacks of mass extinction, the Earth has been an ark for biodiversity for hundreds of millions of years, nurturing the major animal groups that are alive today, from sponges to sea urchins, to animals with backbones. And although the geological record shows evidence of five Phanerozoic mass extinction events, at no point has a major animal group—a phylum—been deleted from nature's inventory of life. This has been pivotal to the success of the biosphere, allowing nature to restore and recolonize from the greatest range of types of body plan. So how might the current human-driven biodiversity loss presage a change that is worse than these earlier extinction events?

Velvet worms are small unobtrusive animals that, unlike their probable marine ancestors, are entirely land-living with a body a few centimetres long, bearing a few tens of pairs of legs.[28] They mostly live within moist forest environments, though some live underground, and they capture their prey by squirting a sticky fluid upon them and hunt everything from spiders to cockroaches. Fewer than 200 species of velvet worm have been documented, but they have wide geographical distribution in the tropics and Southern hemisphere.[29] This small number of species is in stark contrast to the possibly more than 5 million insect species recognized, or even the 100,000 types of tiny unicellu-lar diatoms. In a sense, velvet worms are island species too, each living in a small rainforest habitat. And because of this they are highly vulnerable to deforestation, and to extinction.[30]

Velvet worms are one of nature's special types of body. On the animal 'tree of life' they form a major branch, like that of chor-dates (the group that includes animals with backbones) or mol-luscs (with snails, clams, squid, and others), and like these two groups they have a deep geological history. Cambrian rocks from Canada to China bear fossils of a loosely defined animal group called the lobopods. These were mostly soft-bodied, though some bore small armour-platelets embedded in their skin, and in most fossil deposits it is only these tiny, disarticulated platelets that remain. Occasionally, the whole animal is preserved as a fossil, revealing a worm-like body carried above many pairs of stubby legs. Some of these lobopods bear a strong resemblance to living velvet worms.[31]

Velvet worms have been part of the tree of life and its environments over hundreds of millions of years, and they are the main branch from which other branches—insects, shrimps, and scorpions sprang. No one knows quite how velvet worms have survived past mass extinctions events, because being mostly soft-bodied, their fossil record is very poor. But they have been a success, and long ago made the transition from being a group of seafaring animals to one that lives on land. Humans are fast destroying the environments in which velvet worms live, as the rapid loss of rainforest in South America and elsewhere shows. Not all velvet worms are classed as threatened. For now, some live within protected areas, while others might have the capacity to adapt to human-changed ecologies, like banana plantations. But of all the main branches of life, this is the [32] one that now seems most vulnerable to being lost altogether.[33]

The velvet worms are a message of caution from the biosphere. Their loss from life's tapestry would, from an evolutionary sense, be more final than the extinction of the trilobites, dinosaurs, or ammonites, whose arthropod, vertebrate, and mollusc relatives live on. In the coming decades, keeping a close watch on the fortunes of the velvet worms will tell us a good deal about the planetary severity of the biodiversity crisis that is unfolding.

Restoring great beasts

How might we mitigate the impact of an unfolding biological disaster at the global level, one which might leave poor Mictecacihuatl with no more bones to tend? One good test of the survivability of land-based ecosystems is their ability to support the large and dynamic beasts with which we began this chapter.

One of these beasts is the white rhino, an animal so big that it has no natural predators and whose population size is controlled by the available vegetation where it lives. There are about 20,000 white rhinos in the wilds of Africa, but these are now found only in the south of that continent. The 'northern' species—which occupied savannah regions of central Africa—is essentially gone from the wild. Being up to 4 metres long, white rhinos are the fourth-largest terrestrial mammals, with only the three living species of elephants standing before them. Despite their size, they are not aggressive, and they have suffered terribly from the dual ignominy of being hunted by white colonialists for sport, and now by organized crime rings who kill them for the horns to use in Asian medicines, even though they have no medicinal value. In South Africa's Kruger National Park, white rhinos were extirpated by white hunters more than a century ago. But they were reintroduced in the 1960s, and now number in the thousands, though poachers continue to threaten them. As they graze, a little like gigantic lawn mowers, they have a significant impact on the ecology of the savannah, increasing the diversity of short-grass habitats as a small echo of the megafaunal terraforming of past times.[34] In Kruger National Park the density of the rhinos is still too low for this to provide widespread living places for other animals, but elsewhere, 'rhino-mowing' provides areas for birds, mammals, and insects.

White rhinos are one example of how rewilding through the introduction of a large mammal can have a positive impact on the whole ecology. A similar pattern was seen with the reintroduction of grey wolves to Yellowstone National Park in the USA during the 1990s. They too had been absent from their natural landscape for some 60 years and were returned to the park to help check the growing population of elk. This produced what ecologists call a cascade effect, like the lawn-mowing rhino on the

African plains. The wolves hunted the elk and reduced their numbers, and this curtailed the grazing pressure on willow, aspen, and cottonwood. Willow, which is particularly prevalent along the streams in the park, is very important for the livelihoods of beavers,[35] who use the willow to build their dams. And beaver dams provide small, cool pools for fish, while increased willow provides a habitat for birds.

This is all well and good for national parks, where the density of people is very low and where much of the original ecosystem may still be intact, but such parks represent a small proportion of the Earth's landmass. What happens when wildlife is allowed to develop in more densely populated regions, where it may come into direct conflict with humans? One example of the complex problems that arise from rewilding in places where people also live is the Oostvardersplassen area to the east of Amsterdam. Originally under the sea, the area was drained in the 1950s and 1960s to be used as an area for industry. A rich wetland emerged, but by the 1980s, woodland was taking hold, and the wetlands with their birds were beginning to be threatened. Dutch ecologists planned a cascade effect, hoping that by allowing red deer, horses, and cattle to roam freely and graze, the wetlands would be preserved. But this was an imperfect ecosystem that lacked large predators like wolves, and the numbers of grazers increased dramatically, devastating the plant and bird populations. And because the area is fenced in to prevent conflict with neighbouring farmers, the animals were not able to migrate to new pasture.[36] Trapped in a landscape where their numbers increased dramatically, as did their pressure on the plants in the reserve, all came to a head in the severe winter of 2018, when many animals starved to death. The proximity of this tragedy to a large urban population caused widespread anger at the suffering of the emaciated horses, deer, and cattle. Now the number of grazers in the reserve is carefully managed, but by people and not by natural predators like wolves. And the future development of these reserves may involve connecting several such areas through western Europe, so that the animals can migrate. Oostvardersplassen shows that reintroductions of wildlife into areas that are densely populated by humans take time to get right and require the tolerance and engagement of the local people when they go wrong.

A question of scale

This tolerance of, and support for, wildlife can extend to much smaller scales too, into the potential wild spaces of our farms and individual homes, where for decades we have worked to obliterate the 'pests' that eat our crops and garden plants. In any garden centre or hardware store there will be shelves full of insect-killing chemicals in colourful and appealing plastic bottles. These will carry a warning not to inhale or drink the contents, but no warning to say that the impact on insect biodiversity of your garden may be catastrophic. Something of the order of two-fifths of the world's insect species may be threatened with extinction within a few decades; they are being widely exterminated across both urban and agricultural landscapes, and are decimated by pollution in aquatic settings.[37] There may be more than 5 million species of insects,[38] and their origins go back 400 million years to the Devonian Period. They serve many complex roles in ecosystems and, having co-evolved with flowers for well over 100 million years, they are important pollinators. Bees alone are one-third of all pollinators of flowers. But insects from dragonflies to beetles have many other functions too, such as controlling the numbers of insects like mosquitoes, while termites are essential for breaking down and recycling the many nutrients locked up in fallen trees. Some such relationships have evolved remarkable complexity, and certain termites are farmers, growing their own cultivated fungi to help decay plant matter.

Many insects also provide an important food supply to animals like bats, birds, and even fish. The North American mosquitofish has even been used in pest control. But where these fish have been introduced as a non-native species, they have had damaging impacts on wildlife, not least because they eat a range of other insects as well as mosquitoes. Because insects are deeply embedded in the functioning of Earth's ecosystems, a major loss to their numbers and diversity would have incalculable effects; indeed, would likely cause wholesale collapse to ecosystems, including those that sustain us.

All these ecosystems are essentially pathways for energy—and matters of energy are also central in our disruption of them, as we explore next.

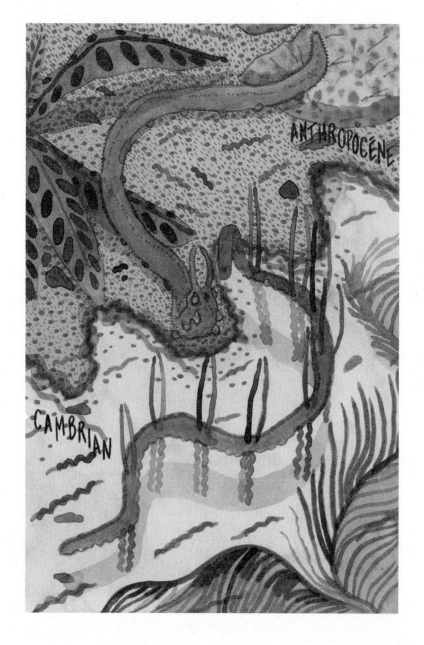

5

A BONFIRE LIKE NO OTHER

For billions of years the energy of sunlight has sustained life on Earth, keeping its surface warm enough for liquid water to exist and, once organisms learned the trick of photosynthesis, providing most of the energy to sustain life. Even after travelling across 152 million kilometres of space, the Sun delivers an average of 340 watts of energy (that is, 340 joules per second) for every square metre at the top of the Earth's atmosphere[1] —enough to power the desktop computer and monitor you may be reading this book on. Some of this energy is reflected back to space, but nearly three-quarters is absorbed by the atmosphere and Earth's surface. The energy received this way adds up to nearly 4 million exajoules a year. And just one exajoule of energy is equivalent to one quintillion joules or, to put it another way, one million million million joules.

The power of the incoming sunlight is converted into chemical energy by plants and stored in their bodies. The chemical pathways the plant uses to capture the Sun's energy are not very efficient: some of the light is scattered, and some wavelengths can't be used, so in the end no more than about 6 per cent of the energy is captured as light falls on a leaf.[2] Many plants also only grow for part of the year and, beyond the tropics, much of the sunlight of winter and spring is lost, while sometimes there is simply too much sunlight to use.

Despite the apparent inefficiency of plants in capturing the Sun's energy, they far outweigh all other life, representing four-fifths of all the biological mass on Earth, with most of this on the land, as forests.[3] They form the basis of the food supply on land and in the oceans too, but in the latter they are represented by countless tiny phytoplankton, reproduced in vast numbers each day, but equally consumed in vast quantities and only having a fleeting existence. On the land, by contrast, their distant relatives stand as trees for hundreds or thousands of years, to be nibbled at intermittently by grazing animals, and eventually broken down by fungi and microbes.

Plants are also the source of the fossilized sunlight that has driven the Industrial Revolution of the human world. Oil and gas are the remains of countless phytoplankton that rained down on to the seabed tens to hundreds of millions of years ago, to be preserved in sediments, buried over time, and converted to hydrocarbons in a setting like a pressure cooker deep in the Earth's crust. On land, meanwhile, generation upon generation of vegetation has accumulated to form coal deposits since the Carboniferous Period, 300 million years ago.

After plants, in the order of quantity of biological mass, are bacteria, representing about 15 per cent, followed by fungi, archaea, and protists (the latter being a loose group of unicellular organisms that includes amoebae, dinoflagellates, and diatoms). Animals, in all of their wonderful variety of forms, represent just over one-third of one percentage point of this biomass! The Earth is, by all measures, the planet of the plants. Half of the biomass of animals is captured in arthropods, small animals with an external skeleton and jointed limbs. Having your skeleton on the outside means that you can never grow really big—you would collapse under your own weight—but this design has allowed for a remarkable range of shapes,[4] from spider crabs to feathery butterflies. Fish come next in animal biomass, followed by worms, molluscs, and jellyfish. Wild mammals and birds are near the bottom of the list, even behind obscure groups like nematodes[5]—and now well behind the mass of humans and their domesticated animals.

The energy of sunlight captured in plants is dispersed through Earth's myriad organisms, beginning with its primary consumers. These may be tiny zooplankton like shrimp in the seas, or large grazing animals like deer on the land. The shrimp may then be eaten by a fish, who is a secondary consumer, or the deer hunted by wolves. And in turn, the small fish might be gobbled up by a larger fish, as a tertiary consumer. At each stage in this food chain—right up to the apex predator like a lion or great white shark—about 90 per cent of the energy is lost. That means that a deer, or a cow, is already a very inefficient way of harvesting the primary energy that arrived as sunlight.

Whatever its place in the food chain, when an organism dies it is recycled, decomposed by bacteria and fungi so that its nutrients can be made available to plants growing in the soil or oceans, and so the cycle begins again. Thus, the biosphere, continuously powered by sunlight, and able to obtain or recycle all the materials it needs to sustain itself, has survived for billions of years.

Below the surface of the Earth is another 'deep biosphere'. Much of this is microbial, living on chemical stores of energy within the minerals of the Earth's crust and supplied with water that seeps into the rock pores where these organisms live. Life in this deep biosphere is energy-poor, but even here the microbes are joined by hardy invertebrates like nematodes and arthropods.

From these energy supplies, mostly sunlight and in much smaller part chemical, something of the order of 10 million species thrive, a diversity of life that has taken billions of years to evolve, and whose species numbers seem to have increased overall, despite periodic mass extinctions. It has taken quite a few centuries to begin to document this diversity. One of the species thus classified, *Homo sapiens*, has become truly remarkable for the amount of energy it now seizes from the Earth.

A bonfire like no other

What would a taxonomist decide as the characteristic of humans, of the genus *Homo* and its one living and several fossil species, that differentiates us from all other monkeys and apes, and indeed from all other species in the biosphere? Aristotle would have said it is our capacity for rational thought. Others have suggested it is our ability to make tools and shape the environment around us, and yet others have emphasized the use of language and writing to communicate information across generations, leading to a diversity of human cultures. Or is it our large brains and cognitive fluidity, or our skeletal anatomy? People have been trying to define what it is to be human since the great Swedish zoologist Carl Linnaeus first gave us the name *Homo sapiens* in 1758, but uncharacteristically for him, he forgot to write a 'diagnosis'[6] for either the genus—*Homo*, or species—*sapiens* (the Latin translates as 'wise man'). Linnaeus might be forgiven for that oversight. When first named, *Homo sapiens* seemed to stand in splendid isolation from the other creatures of this Earth and only much later were other, fossil, species of *Homo* discovered.

So what might be the diagnosis for humans, for those living and fossil? Technology, or something related to it, is not restricted to us. Many kinds of animals have been observed to build shelters and use tools. Birds, ants, termites, wasps, octopi, and even protozoans (in the form of the 'agglutinated' foraminifera, that build their microscopic shells out of sediment grains) can all build nests or protective structures, by selecting and arranging building material from the local environment. These behaviours are to a large extent 'built-in', and under strong genetic control, though the execution of the genetically preordained commands often needs a good deal of behavioural flexibility in practice, for instance in foraging for and choosing the right kind of nest-building material. Physical objects can be used by a variety of animals, as in sea otters that use a large pebble as an anvil to break open shells, dolphins that use sponges to protect their delicate snout when foraging among sea floor sediments, and crows that use twigs to catch and impale insect larvae. In some of these examples, there is modification of a genetically founded behavioural plan by more local factors. With the dolphins, the behaviour seems to be passed on as a learnt culture, particularly from mother to daughter. And, the New Caledonia crow can shape the twigs (or pieces of metal wire) it uses, giving them a reputation for cognitive improvisation.

So, a capacity for manipulation of the surrounding environment is widespread among animals. Nevertheless, humans clearly excel at it, and while *Homo sapiens* is a prime example, *Homo neanderthalensis* was also a technologically proficient species. Who knows what might have emerged from them if their existence as a species had not been cut short. Technology can take a long time to lift off from simple beginnings, as we shall see. For our taxonomic diagnosis of what it means to be 'human', tool use, then, cannot be a defining characteristic. Neither are humans defined by their ability to modify the environment around them, because all organisms do this, and some, like cyanobacteria, have had a profound impact on remodelling the Earth. Culture is not unique either, as is evident in the tool use of different groups of chimpanzees. It cannot be our anatomy, because the differences between ancestral humans and their antecedents are so slight that some scientists continue to argue about exactly when the first species of *Homo* appears in the fossil record. Neither are humans defined by their large brains, cognitive fluidity, and capacity for rational thought, because when the first ancestral humans do appear, in rocks about 2.8 million years ago, they had brains only marginally bigger than those of chimpanzees.

Perhaps the diagnosis of human, of the genus *Homo*, should include some or all of the characteristics that have been mentioned. These led us to develop what is the most striking characteristic of our species of human, *Homo sapiens*. At first slowly and at different rates in different places, then in a great rush, still in full flow, we have become the energy-grabbing species par excellence, our greed for energy now being unmatched in all of planetary history. The diagnosis should therefore include this basic characteristic of greed—perhaps Linnaeus should have named us *Homo avidus*. We are a singular species in terms of our ability to command so much of the Earth's energy resources. By some estimates, each year we extract two-fifths of all of the energy available in the bodies of plants visible above ground. If we were to set this heap of vegetation alight in one gigantic bonfire it would yield more than half of all the energy we already extract annually from oil, coal, and gas. To this energy we need to add that supplied—when we eat them—by our domestic animals, like cattle, sheep, and pigs. We and they together now account for nearly all of the land mammal mass. These extraordinary figures are almost certainly unique in terms of the dominance of one species in the 4 billion years of biological evolution. Such an extreme course is not, of course, without its dangers. The final line of our own diagnosis might turn out to be, 'The only species in history to knowingly cause its own extinction', which (among other effects) would leave no one left to read the diagnosis.

As *Homo sapiens*, the last, currently hyper-abundant, representative of the genus *Homo*, has learnt to dominate so much of the Earth's stored energy resources, some particular tricks helped greatly with this takeover: feeding the plants, for instance.

Bread from thin air

There is a correspondence between the number of people alive and the amount of food available to feed them. For hunter-gatherer lifestyles the Earth seems able to sustain about 10 million humans,[7] but as people developed agriculture and domesticated animals, a surplus of food was produced that could support many more. Steadily, as agricultural and stock-breeding methods improved, the numbers of people grew until at the end of the nineteenth century there were about 1.6 billion of us. Through most of history the chief supply of fertilizer for farming was sourced from natural materials like muck and manure. Each of these is rich in nitrogen and other nutrients like potassium and phosphorus. But as human population grew beyond a billion, the demand for these materials was ever increasing. They were hard to come by, though, and people went literally to the ends of the Earth to extract resources such as guano and 'Chilean Saltpeter'. Then, the doorway to a cornucopia was opened, as people learned to manipulate the atmosphere to make fertilizer.

Nitrogen is a key component of fertilizer. It also makes up the bulk of the atmosphere, a little more than 78 per cent by volume. But, unlike oxygen which is highly reactive with minerals at the surface of the Earth, nitrogen is inert, accumulates over long timescales, and is therefore difficult to capture in a form that can be used by plants. It was only distinguished as a major component of air in the late eighteenth century, by Scottish scientist Daniel Rutherford. Rutherford did not use the term 'nitrogen', but instead referred to his discovery as 'burnt air'[8] in which nothing could live—a quality that was tested on some unfortunate mice. Only in 1790 was the name 'nitrogen' used, by the French chemist Jean Antoine Claude Chaptal.[9] Over the following century nitrogen was identified in rainfall, was shown to be concentrated by certain plants, and was identified as essential for plant growth. It was also discovered that decaying plant matter releases nitrogen, proving the existence of a nitrogen cycle between air, Earth, and life—just as Vladimir Vernadsky would later elaborate on in his development of the idea of a biosphere. Amongst the nitrogen discoveries of the nineteenth century was that of German scientist Johann Wolfgang Döbereiner who in 1823 synthesized ammonia by combining nitrogen and hydrogen using a platinum catalyst. Döbereiner's technique was prescient, though too inefficient to replicate on a large scale. Two other German chemists, Carl Bosch and Fritz Haber, with help from the British scientist Robert Le Rossignol, would perfect the means to do this.

The Haber-Bosch process—as it came to be called—is credited with feeding about half of all the people alive today and is responsible for about half the nitrogen present in your body. It takes atmospheric nitrogen and converts it into ammonia as a prequel to making fertilizer. Haber and Le Rossignol first demonstrated their chemical process in 1909. It combined atmospheric nitrogen with hydrogen from methane to make ammonia by passing these gases over a metal catalyst[10] and condensing the ammonia into small droplets. The German chemicals industry quickly realized the importance of this advance, and under the direction of Carl Bosch scaled it to industrial level, the first factory opening production in 1913. From the beginning, the process was very energy intensive, requiring high temperatures and pressures. Bosch and Haber led distinguished scientific careers, and later received Nobel prizes for their work. Such a monumental advance should have placed them unequivocally amongst the pantheon of great scientists, but both had the misfortune to become caught up in different ways in the tempestuous politics of Europe of the time, while their ammonia-fixing process also provided chemicals for the production of explosives. This opened up another major source of energy, albeit one directed towards destruction.

Haber went on to lead Germany's First World War effort in the development of poisonous gases for trench warfare, with tragic consequences that spread beyond the direct gas victims. His first wife Clara Immerwahr was a highly capable scientist in her own right, and the first woman to receive a doctorate in chemistry in Germany. Clara committed suicide a week after her husband's involvement in the deployment of chlorine gas during the second battle of Ypres in 1915. Whether she took her life in repugnance at her husband's actions, or because of the desperation she felt at the loss of her career, has never been determined.[11] Marginalized in the 1930s for his Jewish heritage, Haber died in Switzerland en route to Palestine where he had hoped to revive his career. Haber himself was reticent about his work. In the address he made on receipt of his Nobel prize 'for the synthesis of ammonia from its elements', he made no mention of the use of the process for making explosives, or of his work on poison gas development.[12] Bosch too was damaged by this era of political turmoil and also became increasingly marginalized because of his repudiation of Nazi policies in the 1930s. Le Rossignol, incidentally, received little credit for his early work on the Haber-Bosch process (though Haber had clearly acknowledged him in the patent).

In some ways the extraction of nitrogen from air by the Haber-Bosch process resembles the evolutionary leap forward that led to oxygen-releasing photosynthesis by cyanobacteria deep in Precambrian times. Both processes used ubiquitous materials to increase the available energy in the mass of plant tissues. But the Haber-Bosch process differs from photosynthesis in one chief respect. To make ammonia it doesn't use the Sun's renewable energy, but the finite resource of fossil fuels, alone being responsible for about 1–2 per cent of all of the energy expended by humans each year. Some of this artificially fixed nitrogen escapes from soils, to over-feed and hence devastate nearby ecosystems in rivers, lakes, and the sea.

The Haber-Bosch process, for all that, expanded the energy available from plants—and this provided the calories that lit the fuse for a human explosion of consumption.

The Green Revolution

By the 1940s, and despite the impact of a global war, the human population continued to grow and at the end of that decade exceeded 2.5 billion people. It was during this decade that high-producing forms of maize and wheat began to be developed as a means to enhance food security, and together with new varieties of rice, these would help to avert starvation, and feed a growing human population. Initially, this work began with the idea of countering the impacts of parasitic fungi such as 'rust', which attack wheat and many other plants, and which had plagued crops since prehistory. The Romans felt the effects of these fungal infestations and celebrated the festival of Robigalia towards the end of April each year, invoking the god Robigus to intervene and preserve the crops.

Mexico had suffered two severe blights from a form of rust in 1939 and 1941 which threatened its production of wheat. Out of this was born the international maize and wheat improvement centre of Mexico, and here began a Green Revolution that would spread throughout the world.[13] One of the scientists of this Mexican institute was the American Norman Borlaug. He led the development of new strains of wheat that were resistant to rust and were also high-yielding, and these would eventually be grown in many other parts of the world. He developed dwarf varieties with shorter stems, that were particularly suitable to poor soils where artificial fertilizers were used (non-dwarf varieties would grow so rapidly that they would topple over under the weight of their own grain). There were other centres of the Green Revolution too, and during the 1960s the Philippines became a focal point for improved varieties of rice that were adopted throughout South East and South Asia.

The Green Revolution's combination of new crop strains, fertilizer, mechanization, pesticides, and irrigation had a dramatic impact on what farmers could grow. In 1961 about 1.87 tonnes of rice could be produced per hectare of land, but by 1980 this had risen to about 2.75 tonnes and is now approaching 5 tonnes. There have been similar increases in the yields of maize and wheat, and in virtually every staple crop that is used by humans, with potatoes and bananas giving the highest yield with over 20 tonnes per hectare. This impressive feat of technology has fed many people, and it also means that compared to the 1960s only 30 per cent of the same area of arable land is now needed to generate the same produce.[14] But since 1960 the human population has more than doubled, and much of the cereal crop is used to feed the animals that we eat, especially cattle, or is manufactured into biofuels. So overall, the amount of land used for crops has risen.

Borlaug's work and those of his colleagues are credited with keeping alive about one billion of the Earth's human population, and in 1970 he received a Nobel prize for his work. He knew then that the crop technologies of the Green Revolution were only part of the solution to growing human consumption, and that wider issues would need to be resolved, including inequality, to avert food insecurity over the longer term. The Green Revolution also relied on the extensive use of human-made fertilizers, of monoculture cropping, and of irrigation, and these would have unforeseen and negative consequences for life.

The increasing use of fertilizer, including those produced by the Haber-Bosch process, led to a doubling of the amount of nitrogen compounds and a trebling of the amount of phosphorus compounds active at the Earth's surface, and this has left a chemical signal that is detectable in the sediments of lakes and seas. It will be a long-term geological marker of human activity. Much of the nitrogen—80 per cent of it—and the phosphorus is simply washed away from the land and finds its way into inland waterways and then into the sea. In bodies of water the nitrogen and phosphorus continue to act as a fertilizer, increasing the abundance of tiny plants and photosynthesizing microbes such as cyanobacteria. These clog the surface waters, forming a green scum that prevents sunlight from penetrating. As the cyanobacteria die and sink, they decay, using up all of the dissolved oxygen in the water. A dead zone results, where little or no oxygen remains until surface currents and wind are able to mix in oxygen again.

One of these dead zones occurs in the Gulf of Mexico. Each year, by late summer, this can grow to cover an area of seabed up to 22,000 square kilometres—an area larger than the country of Wales. The hyper-fertilization is from the inflowing Mississippi River, whose drainage basin covers two-fifths of the interior of the USA, including a vast area of agricultural land. An even larger dead zone forms in the Baltic Sea, covering an area of more than 50,000 square kilometres—greater than the land surface of Denmark. These dead zones spread across the seabed, killing most of the animals living there, such as crabs and fish that need a ready supply of oxygen to breathe. Jellyfish, though, can thrive. Their bodies don't have complex organs that need a strong oxygen supply, and so they bloom in countless numbers in the dead zones.

There are other side effects of the use of nitrogen fertilizers. Only a small part of the nitrogen actually gets to the plants. Some of it is converted to nitrous oxide and escapes to the atmosphere where it acts as a potent greenhouse gas, 300 times more powerful molecule-for-molecule than carbon dioxide. Nitrous oxide also forms from the decay of cattle manure, so that both cropland and pasture can be a source of this gas. And it poses a further threat, because when it reaches the stratosphere it decomposes the ozone that forms a protective layer that stops ultraviolet radiation damaging life at the surface of our planet.

The Green Revolution also relies on irrigation, which drains much water out of rivers, hampering peoples' use of water downstream. On a global scale, irrigation contributes a significant amount of humanity's unsustainable demand for water, which includes about 4,000 cubic kilometres of freshwater per year,[15] or the equivalent of draining Loch Ness dry about every 16 hours.

Although it has been a great success in feeding people and improving livelihoods, the Green Revolution nevertheless exposes our capacity to influence the global chemical cycles of the Earth, and our limited understanding of how these cycles are tied to the long-term sustainability of life. A very tangible example of that limited understanding is the way that we assemble our technologies with little reflection on how their component materials are sustained, or how those materials may be recycled for subsequent use.

Sustaining consumption on a finite world

Nikolai Semenovich Kardashev was a visionary Russian astrophysicist. His family, like many closely connected to the Communist Party in Stalinist Russia, suffered in the purges of the late 1930s; his father was killed and his mother was sent to the Gulag. Taken in by his maternal aunt, Nikolai began to shine as a thinker, and developed a strong interest in astronomy.[16] This led to a career that would allow him to probe the distant reaches of space, both with his mind and through the instruments he helped design. It was his enthusiasm for deep space that led him early in his career to suggest how we might detect the radio signals of extraterrestrial civilizations, and indeed to suggest that some of the signals already detected might be telltale signs of those aliens.[17] In the early 1960s Kardashev was already aware of the rapid year-on-year rise in human energy consumption and surmised that, with its present trajectory, humanity would in about 5,800 years be able to use all the energy present in our galaxy.

In that bold vision of humanity's future, Kardashev went on to describe three types of civilization that might be detectable from radio waves in our universe. A Type 1 civilization would be able to utilize all the energy falling to the surface of its planet from its local star, and Kardashev thought that humanity was approaching this level of energy utilization. A Type 2 civilization would be capable of harnessing *all* the energy of its local star, and a Type 3 civilization could harness all the energy of its local galaxy. Kardashev thought the possibility of detecting Type 1 civilizations would be extremely low, but that Types 2 and 3 held promise, even with the technology available in the 1960s. His vision was a catalyst for many programmes to search for interstellar and intergalactic civilizations. But, brilliant as his lateral thinking was, it did not consider the problems posed by the non-renewable sources of energy needed to power civilizations to Type 1, and that these would inevitably run out, but not before they had caused widespread damage to air, water, and life on habitable planets—and, that these levels of damage could derail the possibility of sustaining a technologically advanced civilization over time. To arrive at a fully Type 1 civilization, humans still have some way to travel, including converting the bulk of their energy systems to using the power of sunlight, sustainably, just as photosynthetic microbes learned to do in the deep Precambrian.

By the early twenty-first century human energy consumption had grown to staggering levels of over 950 exajoules per year. Over half of that energy[18] is from coal, gas, oil, and nuclear, whilst the other part is taken from our use of vegetation,[19] some for food, but most to feed the animals we feed on, and some simply to be burned. By comparison, all the energy stored in all land plants on Earth, both the leaves and twigs above ground and the roots below, is about 2,190 exajoules. Those trees and plants would make a very big bonfire; humanity's own fire would currently reach nearly half as high. Perhaps Kardashev-type civilizations are not detected in the universe because they are simply too greedy for resources and burn themselves out.

How might we avoid a 'burn-out' on planet Earth? A simple answer is to find a source of energy that does not cause destruction of the environment. Kardashev offered no real clue to this, and instead could see no reason why the rate of energy consumption would not continue to rise. He was writing at a time of unprecedented growth in the world economy and before environmental issues had come to the fore, and also when space research was in its nascent days of flowering, with all of its possibilities for discovery of new worlds. But one such obvious source of energy is sunlight, as this powers the biosphere. If just a tiny fraction of the energy of sunlight could be trapped in solar panels with a 10 per cent efficiency and covering much less than 1 per cent of the Earth's surface, it would be more than enough to replace our dependence on fossil fuels.[20] That sounds like a quick and easy fix, so why aren't we doing it?

An immediate problem is that solar panels themselves are assembled from many components sourced from different places. They use a thin layer of a material that generates electricity when struck by sunlight, such as silicon, and for ultra-thin applications, combinations of metals like cadmium, tellurium, indium, selenium, and gallium. Many of these metals are either very rare or difficult to obtain. And in the case of indium and gallium they are recovered by the energy-intensive smelting of ores of zinc and aluminium. For the latter it takes about 200 megajoules of energy to recover every 1 kilogram of the metal, and about 50 megajoules for zinc (one megajoule is a million joules).[21] And there is still the process of extracting the gallium and indium. For every megajoule of energy expended, about 100 to 500 grammes of carbon dioxide is added to the atmosphere.[22] So by the time the solar panel is fitted, a lot of energy has already been used and greenhouse gas added to the atmosphere.

Making solar cells already embeds a considerable amount of energy in their manufacture, but there is also a second problem of storing the electricity to use later. The only way to do this is via batteries, but these too will consume even more raw materials, including yet more tellurium, and other metals like cobalt. These are now being sought from the deep sea, on seamounts and adjacent deep-sea vents with their delicate, unique, and diverse ecosystems.[23] It is the same technology that is being touted as a solution to climate change, to make you think you can drive your electric car without damaging the environment.

To capture even a fraction of the incoming energy of our star would power all our energy requirements long into the future. But doing that without major damage to the environment still remains an unsolved problem. Unless we can grow our solar panels from the soil beneath us, nurture and give them life and allow them to return to the soil at the end of their lives, we will continue to have this problem. Before we can inhabit this planet sustainably with the rest of its life, we need to navigate the immediate problem of how to re-embed ourselves and our devices back into nature so that we don't use up all of the Earth's resources without giving something back. Without that we will, in the Kardashev classification, simply be part of the profound silence of space.

Growing the sustainable cities of the future

A dark vision of a future world in which humanity's problematic relationships with nature remain unsolved was shown in Fritz Lang's classic 1927 film *Metropolis*. It was set a century in the future, in the year 2030, just a few years hence from now. We are not told how many people live in Metropolis, but the city covers a million acres, and its gigantic skyscrapers were a futuristic glimpse of the megacities that were to come. Metropolis seems a hostile place for most of its working inhabitants, ruled over by an indifferent elite. And, it appears, a hostile place with little space for nature too. Fritz Lang did not indicate what lies beyond the city or how Metropolis lived alongside the natural world, but its dark cityscapes suggest that this was not a benign relationship. A new Metropolis emerged in the Los Angeles of Ridley Scott's 1982 film *Blade Runner*. It too was set in a dystopian future, in the Los Angeles of 2019, where its 119 million inhabitants lived in perpetual gloom and rainfall. Los Angeles has not grown to be quite the monstrous city of the film, but some future cities are projected to have populations of nearly 90 million by the end of the twenty-first century. These sombre visions of near-future *Metropolis* and *Blade Runner* are joined by the diametrically opposing visions of happy people and gleaming skyscrapers in Star Trek's San Francisco headquarters in the twenty-third century. So which of these future cities is more realistic?

In 2007 we *Homo sapiens* became a predominantly urban species for the first time in our 300,000-year history. In that year there were 6.6 billion of us alive, with over 3.3 billion of us in villages, towns, and cities. The global population has grown to nearly 8 billion people as we write in 2021, with over half of us urban, and 1.7 billion of us living in large cities, those with more than one million people. Many of these cities are growing very fast. Some are projected to become giants. Greater Tokyo already covers three times the area of Metropolis. Any passing starship from a Kardashev Type 2 or 3 civilization would have noticed the dark side of the Earth ablaze with the lights of over 500 cities with more than one million people, and they might have surmised that we were approaching a Kardashev Type 1 state. From analysing the composition of our atmosphere, they might also have surmised that our energy resources were still non-renewable, that we were not yet sufficiently developed to gather the energy of our local star efficiently—and that the next couple of centuries would make or break us. Perhaps, in strict intergalactic protocols, they chose not to intervene until we have reached that sustainable state.

Cities are chief among the consumers of non-renewable forms of energy that threaten our continued existence and that of the biosphere. City dwellers, on the whole, consume much more energy than rural humans. Detached from the landscapes that feed them, cities have grown to be gigantic parasites, consuming energy at huge and increasing rates, and returning only pollutants—like carbon dioxide, plastics, and pesticides—to the environment around them. Some of the projections for the future energy consumption of cities are truly alarming, and if unchecked, suggest that by the later part of the twenty-first century we would be burning through the equivalent of over 17,000 million tonnes of oil a year.[24] Clearly, a futuristic city like Metropolis, or the gigantic Los Angeles of *Blade Runner*, would not survive long. But other paths might, perhaps, be opening to allow cities to become genuinely sustainable.

For, there are precedents for humans to build *with*, and not against nature. In the Cherrapunji region of India's northern province of Meghalaya, the indigenous Khasi people build wooden bridges that stand for hundreds of years. These are made by gradually guiding the roots of the Indian rubber fig tree, *Ficus elastica*, across river ravines, by strapping its roots to tree trunks. As the roots grow, the bridge gets stronger, and its lifespan is only limited by that of the trees that support it.

If city developers followed the bridge builders of Meghalaya, they might have at their disposal some of the most dynamic organisms on Earth: microbes, that have had billions of years of practice at designing and contributing to sustainable ecosystems that can include large, robust physical structures. They can form biofilms that offer protection from dehydration and attack from antimicrobial materials, and these structures can accumulate thin layers of limestone. Microbes learned this technique far back in the Precambrian and used it to build the first large biological structures on Earth, the stromatolites. Biofilms can now be manipulated to create self-healing buildings that fill in cracks or stabilize foundations. And bacteria can be manipulated to make certain shapes and sizes of crystals of limestone, and to grow structures that could use carbon dioxide extracted from the atmosphere.[25] Perhaps in this way the longevity and recyclability of buildings might be enhanced, and the energy needed to make them would be used more efficiently.

For now, the idea of biologically grown buildings is a nascent technology that will take time to develop. But there are many other ways that cities and communities can live more beneficially with the life around them and learn to consume less voraciously. This can be as simple as building with sustainable sources of wood that capture carbon in their structure. We can recycle materials within the city, to make new bricks and concrete, just as a natural ecosystem would, rather than bulldozing and building again.[26,27] This might help to quell our excessive use of Earth's materials, including the limestone and mud that goes into making cement, of which we use 4,000 million tonnes each year.

Our towns and cities are also responsible for much of the freshwater used by humans each year, and this is an energy-intensive process.[28] We transport water over long distances, capturing it from nearly half of the land's surface,[29] and then we treat it with chemicals before using it. And yet water that falls within our local neighbourhoods is often wasted, funnelled away from the city along streets and drains. Often this water becomes contaminated with pollutants washed from city roofs, and causes damage to river and lake ecologies downstream. As we concrete over our driveways and dig up the grassy borders of our streets and homes we exacerbate this problem, just as we encourage rivers to flood towns by cutting down the vegetation in the landscape around them. But we could conserve the water that arrives in our cities, just as a natural ecology would do. We can build streets with borders of green that absorb the rainfall, wetland habitats where the water can collect, and infrastructure that captures the water to replenish the local water table. In our homes we can recycle water that is euphemistically referred to as grey, yellow, and brown. These multi-coloured liquids that we flush away without a second thought contain materials that are useful to the environment, including phosphorus and nitrogen that are fertilizers. No natural ecology would be so wasteful.

And what about the energy consumption of cities? While natural ecologies source much of their primary energy from the Sun, as we have seen, cities draw in enormous supplies of energy across long distances, and from non-renewable sources. And as a result, cities are not sustainable in their present form. They consume the equivalent energy of 10,000 million tonnes of oil each year, and this is set to grow. Even taking the capacity of the largest oil tankers, that's the equivalent of 20,000 of them, lined up in one year to deliver this energy. As a result, cities are responsible for three-quarters of the carbon dioxide that accumulates in the air each year, that is causing a rapid warming of our climate. How can this unquenchable thirst be sustained? Some cities have begun by installing lighting systems that are low energy and can learn to switch off at times of low traffic. Such simple interventions need the support of people too, whose individual patterns of human consumption—how we travel, the white goods we buy, and how we use and recycle them—are even more important solutions to this problem. An electric car still damages the environment, no matter what the TV commercials tell you. An integrated public transport system will do this also, but at a lower level than individuals driving to work in cars.

Just beyond the horizon, and like self-healing buildings, new technologies might provide some answers to our energy problems. One company in Holland led by Marjolein Helder and Nanda Heshof is trying to exploit the natural energy of plant matter in soil, and their technology can already power small devices like LEDs and WiFi hotspots.[30] This process does not kill the plants but taps into the excess matter they excrete via their roots, and which is decayed by microbes, releasing energy. It might allow farmers growing crops to simultaneously make energy to light their homes at night. Other researchers are trying to devise new ways of harvesting the energy of the sun by mimicking the structure of leaves and the biological processes of plants.[31] If this was eventually to lead to cities that grow their own power supplies, just as a natural ecology does, then finally we would be approaching fully sustainable cities.

Biologically built buildings, electricity from soil, sunlight trapped by solar cells that function like leaves. These are nascent technologies. Yet, like lightbulb pioneer James Bowman Lindsay's prediction in the 1830s that one day our houses and industry would be powered by electricity, they may be a glimpse of a better future. Currently, though, despite the best efforts of some individuals and communities, our patterns of consumption remain akin to a parasite on the biosphere. Diagnosing this condition, as we explore further in the next chapter, is straightforward; finding remedies is rather less so.

6

THE BITE IN YOUR HAMBURGER

In 1948, the McDonald brothers, Richard and Maurice, made a decision. Their California restaurant, 'McDonald's Bar-B-Que', offered a varied menu—but was making most of its profit from hamburgers. So, they rebuilt their business around a drastically diminished menu, selling just hamburgers (with or without cheese), fries, coffee, and soft drinks. The customers were encouraged to eat this most basic of meals quickly, in a restaurant designed to be cold, the seats uncomfortable and unsociably arranged, and the drinks cartons cone-shaped so they could not be put down. Having eaten quickly, they left quickly, to make room for more customers. A recipe for indigestion, and business failure soon to follow? Extraordinarily, that was not what happened (though the indigestion may well have featured from time to time). More and more customers piled into what had just become 'McDonald's'. The idea grew, and in 1953, the first franchises began to spread the concept more widely. It established a model that would lead to an explosive growth of burger restaurants across the world.

Just what is it about hamburgers? The name itself derives not from ham (as hamburgers are made from beef) but the town of Hamburg. But which one—the city in northern Germany or that in New York State? Both have separate claims as the source of burgers, while something akin to a hamburger was already being consumed in Germany as frikadelle, much earlier than the supposed late-nineteenth-century invention date of the hamburger, and even found its way into Indonesian cuisine—via the Dutch—as the meat and potato fritter called perkedel. All these claims, though, may also be reinventions, as the Roman cookbook the *Apicius* records how to make a patty called 'Isicia Omentata'[1] from minced meat, peppers, pine kernels, and garum—the Roman's fermented fish sauce—and though this cookbook was transcribed in the late fourth or fifth century CE, such a minced-meat patty may have been circulating in the Roman world for some time. Nevertheless, most sources trace the origins of the modern hamburger to the United States, with its classic design of a meat patty sandwiched in a bread bun. One of these origin stories centres on the small town of Fairmount in Indiana, which is also the birthplace of another American icon, James Dean. Here Bill Dolman operated a lunch wagon between 1885 and 1907 that is reputed to have fried the first hamburgers.[2]

Whatever their origins, such burgers were being sold in markets and by street vendors as far apart as Texas and New York State by the start of the twentieth century. Portable and high in calories, they were the perfect food for a burgeoning city population on the march. And as Americans took to their cars in the post–Second World War economic boom, drive-in burger restaurants quickly spread from the west coast to the east. By 2020 there were 39,198 McDonald's restaurants worldwide.[3] They served 25 million people each day, and an estimated 75 burgers a second, or 2.36 billion burgers a year—a large number, but still only a small fraction of total hamburger production. Overall, they may have sold more than 300 billion burgers since the company was founded. The growth of McDonald's was so rapid that when the Earth System scientist Will Steffen was establishing criteria by which to judge the accelerating global changes of the late twentieth century, he chose the spread of McDonald's restaurants as one of his key markers.[4] To the McDonald's tally of restaurants we might add the 18,573 Burger Kings (as of 2020),[5] and the many independent restaurants that sell burgers around the world. It is difficult to estimate the numbers of meat patties eaten each year globally, but in the USA alone this is thought to be about 50 billion.[6]

What area of the Earth's surface would burger restaurants cover if they were lined up back-to-back? Taking an estimate for the average McDonald's or Burger King, as covering a land area of about 400 square metres[7]—averaging for the spacious drive-thru restaurants and the more cramped urban ones—that gives us a total surface area of the order of 23,108,400 m^2 (57,771 restaurants x 400 m^2), or 23 square kilometres. Or, burger restaurants lined-up end-to-end for 1,150 kilometres, the distance from New York to Chicago.

This would be an impressive line of restaurants in itself. But keeping them in work needs a much bigger geographic area, in the form of the land needed to supply them with beef. Because most of the incoming energy of sunlight is lost before it is trapped in the plants that cows feed on, and because 90 per cent of the energy of plants is lost in being converted into cows, it takes about 164 m^2 of land to produce just 100 grammes of beef protein.[8] If 50 billion burgers are eaten in the USA each year—and we use that as a conservative estimate for global burger consumption—and they each weigh about 50 grammes, then that equates to 2.5 million tonnes of patties, requiring a surface area of 4.1 million

THE BITE IN YOUR HAMBURGER · 159

square kilometres of land for the cattle to graze. That's an area of land roughly equivalent to 75 per cent of the Brazilian rainforest, or nearly half the surface area of the United States. All that consumption of land—and that just for one country—has happened in the space of 80 years. But the roots of this business go much further back, to some 11 millennia ago, as the Earth last came out of the depths of an ice age.

Taming the beast

The ancestor of all domesticated cattle today is the auroch, an extinct kind of wild cattle that used to range across much of Europe, Asia, and North Africa. There are eyewitness accounts of what aurochs were like, not least from Julius Caesar, in his accounts of the Gallic Wars. Caesar liked to make observations on natural history, though these were not always to be trusted.[9] Elks, he said, had no kneecaps, and so could be captured by part-way sawing through the trees that they leant against to sleep, and he reported unicorns in the vast Hercynian Forest of what is now Germany. Both are the tallest of tales—but with aurochs, Caesar's memory served him better. These were the size of an elephant, he said (he had in mind the extinct North African elephant, which is somewhat smaller than the Asian elephant living today, and so was not too far off the mark—a bull auroch could be 2 metres tall), extraordinarily fast and strong, and aggressive ('they spare neither man nor wild beast that they have espied').[10] Hunting this fierce wild beast then was one way for a young man to prove his manhood, and the hunting continued over the millennia. The number of aurochs steadily fell, until the last of its kind died (of natural causes, in the case of this individual) in a Polish forest in 1627.

Individual aurochs were more or less untameable, so it could have been no easy task to form the first domestic population of these animals, from which most modern cattle are descended. This took place, as revealed by mitochondrial DNA analyses of modern cattle and ancient bones, combined with archaeological evidence, in a small area of what is now the Euphrates and Tigris valleys of Syria and Turkey, a little under 11,000 years ago.[11] This single event involved the domestication of only some 80 female aurochs, in this small area, hemmed in by the surrounding mountains. It took something like 2,000 years, the successive generations of these animals being tended by many generations of the equivalent of perhaps a couple of small Neolithic villages, before a visibly different and, importantly, a more manageable kind of animal emerged, which then went on to be spread, by migration and trading, across Europe from about 9,000 years ago.

Other experiments in this slow, difficult, and no doubt dangerous experiment in domestication may have been tried by other communities in the region. If so, they have left no genetic marks in today's cattle. Four thousand years ago or more, another auroch population was domesticated, perhaps in Egypt, to produce the zebu cattle now characteristic of South Asia, distinctive through a fatty hump upon its shoulders and tolerant of tropical temperatures.

Cattle, once tamed, gave much to humans: their meat and milk, their strength for ploughing, their skin for leather, their dung for fuel and fertilizer, and at times themselves as a direct form of currency. The last of these most clearly betrays their relationship to humans, one also hidden within the name itself. 'Cattle' is related to the word 'chattel', or possession, and to 'capital' in its economic meaning. An even older Anglo-Saxon word 'feoh' (now preserved within 'vieh', the modern German word for livestock), has similar connotations of property. Cattle, through all that they gave, were therefore wealth.

Humanity is now, in these terms, a billionaire. Estimates of the number of cattle vary, but the Food and Agriculture Organization suggests some 1.5 billion head of cattle now exist on Earth at any one time, ranging from more than 200 million in Brazil, to a little over a dozen in Greenland. To this one may add a billion pigs, a billion sheep, and—their numbers rising most steeply of all these animals over the past few decades—about a billion goats. The scientist Vaclav Smil has spent a lifetime tracking the scale of the human enterprise, including simply, but tellingly, the size and weight of this part of it.[12] At the beginning of the twentieth century, the total weight of all the 1.6 billion humans on Earth was about 65 million tonnes, already outweighing the 50 or so million tonnes of all wild land animals, while both were outweighed by domestic animals living then, at about 175 million tonnes. A century on, this calculus had stretched to proportions that might be regarded as grotesque: in the year 2000, the 6.1 billion people then alive had a mass of some 275 million tonnes, while wild animals had fallen to 25 million tonnes and domestic animals had ballooned to 600 million tonnes. Humans, then, made up a little less than a third of the mass of all land animals, while their domestic beasts bulked up to a little more than two-thirds; wild animals made up just 3 per cent of the total. Now, as we write in 2021, with 7.9 billion humans on Earth, consuming proportionally more cattle, pigs, sheep, and goats, and wildlife in further decline, those wild animals are becoming an even smaller proportion within Earth's re-engineered biosphere.

That is the price that is paid by the biosphere (perhaps not so much by humans, or at least not yet) for an ever-available hamburger. It makes a striking story, but one eclipsed, in some ways, by that of the chicken sandwich.

The Anthropocene chicken

In the forests of South East Asia lives the red jungle fowl. Its existence as a species is not endangered by the activities of humans, and the International Union for Conservation of Nature lists it as of 'least concern'. It is a beautiful bird. The male wears feathers of gold, red, bronze, and iridescent blue and green. He has an impressive array of tail feathers, 14 in all, and upon his head he bears a multi-pronged crown of red. The female's plumage is more modest, but elegant, nonetheless. Her neck is covered with golden feathers above body and tail feathers of brown. In the wild, these highly active birds can live a dozen years or more.

The problems for this bird began about 8,000 years ago, when humans began to domesticate them.[13] Possibly they had been hunting jungle fowl for food much longer than this, and perhaps also stealing their eggs. Over time it seems that the red jungle fowl was domesticated in several different places at several different times, and that it hybridized with other jungle fowl, most notably with the grey species of South Asia, eventually to produce the domestic chicken. The Harappan people of the Indus Valley were living with domestic chickens by 4,000 years ago. By about 2,500 years ago, chickens had spread to Europe.

Scratching around in countless backyards and improvised henhouses, for millennia these domesticated chickens provided eggs, meat, entertainment (of a kind, with cockfighting popular in some societies), and were sometimes used in ritual sacrifice. Their bones are found at many archaeological sites. Different breeds slowly arose, but the basic pattern of chicken anatomy stayed pretty constant. One routinely used measure of the chicken is the length of the tibiotarsus, a lower leg bone that is one of the commonly found and easily recognized parts of the skeleton. A compilation of the dimensions of archaeological tibiotarsus remains from London showed that the domestic chicken legbones of Roman times were similar to those of the wild ancestor. Then, over the following centuries, they edged upwards in size into medieval times, subsequently staying much the same, even well into the Industrial Revolution and early twentieth century. But then came the Chicken-of-Tomorrow programme of the immediate post–Second World War years, and a monster was born.

The Chicken-of-Tomorrow initiative was an intensified breeding programme in the USA, designed to create birds with more meat on them, and that converted feed into weight gain more quickly. Organized as a contest,[14] it was run on scientific lines, taking 100 chicks from each chicken farmer or breeder, and growing them in exactly uniform conditions, and then killing and preparing them in an exactly uniform way after 12 weeks, and weighing, measuring, and judging the resulting carcasses. From this competition two lines were selected for crossbreeding. Already by 1952, it was judged to have precipitated 'a change of gigantic proportions' in the USA. This 'gigantic' change continued over the succeeding decades, to produce a new kind of bird.

These are the broiler chickens that we now see in row upon row of supermarket shelves around the world. They are birds with white feathers, having lost the brilliant plumage of their red jungle fowl ancestors. Broilers are factory reared under artificial light, often in spaces where more than a dozen chickens occupy a square metre of space. This unnatural environment produces many abnormalities in the bird's skeleton, particularly weaknesses in the legs and necks, and many birds suffer from dermatitis because of their close proximity to litter. They grow hyper-rapidly to the point where they are culled in five to seven weeks. By that time they have become four to five times heavier than a chicken from the 1950s, with a skeleton to match: that tibiotarsus bone, for instance, has now roughly doubled in both length and width. It is still the same species biologically, albeit with a genome that has suddenly become much narrower than that of its wild ancestor. But to a palaeontologist it would be a different species, a kind of *Incredible Hulk* of the chicken world—one that has arisen in a few decades.

Its increase in bulk does not give it superpowers, though: rather the reverse. The end of a broiler's short life is a process reminiscent of a medieval torture chamber, where the birds are shackled upside down on a moving conveyor belt, immersed in water where they are stunned by an electric shock, before having their throats cut (the cattle for our burgers and the pigs for our bacon do not have a much easier death). It is an environment far removed from the jungles of South East Asia, and one that we may take care to forget when we reach for a chicken sandwich. It is another example of how we have adapted the biosphere to our needs, and of our well-honed ability to ignore the suffering of other animal species. Even if one was to be rescued from the abattoir, it would likely not live much longer, as the leg and breast muscles have grown too huge for the heart and lungs to cope with. It is a technological construct, albeit a living one, that can only survive (briefly) in technologically controlled conditions. And it has taken over the world.

At present, some 65 billion chickens are raised and eaten by humans each year and 80 million tonnes of eggs are consumed across all ice-free landmasses. There are some 23 billion domesticated chickens alive at any one time.[15] These are eye-wateringly huge figures that far exceed the numbers of any wild bird species: the next most populous bird is the tiny red-billed quelea of Africa, but these small seed-eating birds only manage about 1.5

billion (and their numbers may have burgeoned because they feed on crops that are cultivated by humans, such as wheat, sorghum, and millet). Combining the bulked-up mass of the modern broiler chicken with its (still rapidly growing) numbers means that it alone now outweighs *all* wild birds in the world something like threefold. Its overgrown bones, now discarded into landfills worldwide, are set to become a distinctive fossil of the future, one that can act as a symbol for the human transformation of the Earth in what is coming to be known as the Anthropocene Epoch. It is in many ways an apt symbol for these times, though the changes it represents run much more widely.

Imperiled landscapes

As the climate ameliorated some 11,000 years ago, the first people returning to the British landscape were Mesolithic hunters who brought with them sophisticated stone tool kits, which they used to kill and butcher the animals they followed into the warming landscape. They used wood too, to make substantial structures. At the Mesolithic settlement of Star Carr on the edge of ancient Lake Flixton in the Vale of Pickering, Yorkshire, 9,000 years ago, they built timber-framed dwellings and platforms along the edge of the lake. Star Carr was neither transitory nor makeshift: it was a village. Here there is also evidence of periodic burn-back of vegetation, of an attempt to manage the land. These early settlers may have had only a marginal impact on the forests around them for several thousand years, and only when the agricultural revolution of the Neolithic reached Britain about 6,000 years ago did forests begin to be cleared, as land was converted to pasture or crop farming. These Neolithic people shaped the landscape in other ways, building large monuments across the British Isles from Stonehenge on the Salisbury Plain to Callanish Standing Stones on the Hebridean island of Lewis. Landscape transformation accelerated during the Bronze and Iron Ages and Roman occupation, with the wider use of ploughs. By the time of the Doomsday Book in 1086 CE, perhaps only 15 per cent of England was still covered by woodland. It has never crept above this level to the present day.

Fast forward to the twenty-first century and over 73 per cent of the English landscape is farmed, and over 12 per cent is urban.[16] A little less than 15 per cent is called 'natural', but this is transformed too. As the natural landscape has receded, so too has the wildlife. Large animals like bison disappeared as early as the Neolithic, brown bears in the early medieval period, and the last wolves roamed Scotland in the late seventeenth century. But the effects on the landscape are much more profound than the loss of these visible animals, and now are magnified by climate change, the introduction of non-native species, and the widespread use of pesticides. Even the small animals are declining. A third of wild insect pollinators such as bees and hoverflies have declined over wide areas of the UK in the last few decades.[17] Many woodland bird species are in serious decline: the lesser spotted woodpecker and spotted flycatcher by more than 80 per cent.[18] Mammal species are threatened too, from the hedgehogs whose UK population plummeted from perhaps 30 million in the 1950s to fewer than one million now, to the 200 wildcats that remain in the wildernesses of Scotland.

Elsewhere, landscapes have also been drastically changed by human activities. When Europeans first arrived in North America, much of the landscape was deeply forested, from the deciduous forests in the east, to the coniferous forests of the west, and the boreal forests to the north. In the northern Arctic lay great areas of tundra, and in the central and south-western regions prairies and scrub. First Nation people arriving across the Aleutian archipelago from East Asia towards the end of the Pleistocene made their homes in all of these landscapes, adapting to and interacting with the wildlife as it evolved in the post-glacial world. These interactions almost certainly caused extinctions of many large mammals, but this does not appear to have precipitated a wholesale continent-wide collapse of ecosystems.[19]

Within a few centuries of the arrival of Europeans, much of the cultural diversity of the original Americans, their ways of seeing the world, and ways of doing things, were swept away, and with it much of the original vegetation and biodiversity as intensive agriculture spread across the wilderness, enforced by a patriarchal society with a strong emphasis on land ownership. For First Nation peoples, from the Algonquian of the Atlantic coast, to the Chumash of the California coast, this was a disaster. On Manhattan Island the hickory trees are long since gone. Once the Lenape came here to seek out the trees—manaháhtaan is the place where bows could be found—and the rocky outcrops of Central Park still evoke a vivid sense of the ancient landscape they walked through. To the north, another Algonquian people, the Mohicans, celebrated in James Fenimore Cooper's novel, have been expelled from their homeland. There was no Noah's Ark for these cultures, as the rising tide of Europeans drowned them.

This process of extirpation of whole cultures and ecosystems, and of the assimilation of the land, and sea, is ongoing and is a root cause of the unfolding mass extinction. Some 95 per cent of the global, ice-free landscape is now human-influenced in some way,[20] whilst two-thirds of the oceans show some form of increasing human impact.[21] These are extraordinary figures, and ones that suggest there are few truly wild places left.

The conversion of land for cattle and crops has been going on for millennia. One group of archaeologists[22] has plotted these changes to show the unfolding human influence on the land-scape worldwide, beginning with the development of agriculture in the Middle East over 10,000 years ago, and a little later in East and south-west Asia. By 6,000 years ago nearly half of the land witnessed some form of agriculture. By 2,000 years ago many regions of the Earth showed the footprint of agriculture, mirror-ing the changes evident in the British landscape, and the impact of humans on Earth's terrestrial ecologies had become visible everywhere—and this process would accelerate, as human pop-ulation grew.

The human population is estimated to have reached one billion in the year 1800. That year saw some far-reaching changes. In Eu-rope, the Italian scientist Alessandro Volta introduced his battery to the world. Made from zinc and copper and an electrolyte of sulphuric acid, it is the template for modern batteries. In Vienna's Burgtheater, Ludwig van Beethoven's first symphony premiered, while in the New World, Gabriel Prosser, a Virginia slave, planned a rebellion that, although brutally suppressed, was to herald the fight for emancipation.[23]

It had taken 300,000 years for humans to reach a billion peo-ple. But it would only take another 127 years to add another billion, and another 33 years to add the next. Thereafter a new bil-lion was added to the world population about every 12 to 13 years. The consequences for the Earth are still in train. Over the past two centuries, land-use change has accelerated rapidly, driven by this explosive growth of human population, wealth, and changing food consumption.[24] This leaves little space for nature.

If we asked nearly 8 billion people living on this Earth to stand together, each in one square metre of space, that would take up 8 billion square metres, or 8,000 square kilometres. We could fit everyone on Earth into about half the surface area covered by greater Tokyo, or onto about 130 Manhattan Islands. Viewed this way we might think that there is still a lot of space for people to fill, and that there is no real constraint on how our population can grow. That might work if we obtained our energy by photosynthesis, but for an omnivore with strong carnivorous instincts, such a simple relationship does not follow. For a start, only about 71 per cent of the land is habitable by humans, the rest being desert or covered with ice. That is still a substantial 104 million square kilometres.[25] Of that land, about half is already used for agriculture, whilst the remainder is forest and scrub, and a small part (about 3 per cent) urban. Of the land that is farmed, three-quarters is used for meat and dairy, whilst the remainder is for crops. But dairy and cattle only contribute about one-fifth of the global calories we require, and most of that meat consumption is focused within wealthy countries or is consumed by the wealthiest people in countries. About one-third of the land used for crops is also harvested as animal feed. If all of us adopted the meat-rich diet of the USA (consuming nearly 37 kg of beef a year per person), we would need nearly double the farmland we currently use, and even more if we were to adopt the typical diets of Australia (41 kg) and Argentina (55 kg).[26,27] We would need to cut down most or all of the world's remaining forests to accommodate this, fashioning a world where gibbons and rhinos live only in zoos, and where there would be almost no wild spaces for humans—or other organisms—to escape to. But if we were to live with a typical Indian diet, we would need less than half the farmland we currently use, and we could return huge areas of the landscape to wilderness.

Here, we might look at the amount of land needed to produce different kinds of food. It takes a little less than 190 square metres—the size of a singles tennis court—to produce 100 grammes of lamb protein. Cattle too, need a lot of land, about 164 square metres for each 100 grammes of protein—enough to make two burgers. At the opposite end of the scale, grains like wheat, rice, and maize need just 4.6 square metres, and peas need just 3.4. Somewhere in between is the land needed to produce 100 grammes of milk and cheese, about 27 and 40 square metres respectively, while eggs, or rather the chickens that produce them, need a little fewer than 6. This means that in those countries like Canada, Australia, Argentina, and the USA, which consume more than 30 kg of beef per person a year, just halving this consumption could allow huge areas of the world to be returned to the biosphere. Moreover, in a country like Brazil, where beef consumption is—unusually for the world—growing, it might save great swathes of rainforest.

The problems of our food consumption are complicated further when waste is factored in. Something of the order of one-quarter of all food produced rots, either left in the field, lost in the process of getting it to people, or worse still, tossed into household bins. Here too, there is much variation between countries. Wealthy Europeans and Americans each cast away about 100 kg of food a year, whilst in South East Asia the figure is only about one-tenth of this, though there is much variation within countries between rich and poor people.[28] The amount of food wasted in the world's richest country, the USA, could easily be used to feed the one-fifth of humanity that is undernourished.

Our approach to food is suicidal, or more precisely, given the huge and growing inequalities between rich and poor, homicidal. It is now becoming impossible for all the world to eat like the world's wealthiest nations and wealthiest people—even at the cost of all the Earth's remaining wildlife. Something has to give. Let us say that we try to seek to achieve this equitably, rather than by following the current trajectory of preserving the consumption habits of the wealthy and driving the poor to the wall. Human population is growing towards 9 billion people by the middle of this century. There needs to be enough food for everyone to eat, even with a growing population, while still allowing large areas of the world to be wild and thrive. It is quite a challenge.

But it is not just on land that the biosphere is threatened. Our influence has been rising on the high seas.

Shrinking fish

Over two-thirds of the Earth's habitable surface is the seabed, and above it the ocean's waters are an average of 4 kilometres deep, and also teem with life.

The fish of the sea have a long history that plays out in the fossil record over hundreds of millions of years. The most ancient fish was not much bigger than whitebait, but some would eventually evolve to be enormous, like the 10-metre long *Dunkleosteus* of the Devonian, the even more gigantic 20-metre long *Leedsichthys* of the Jurassic, and the almost as gigantic shark Megalodon— three times the size of a Great White—that went extinct over 3 million years ago. Currently over 34,000 species of fish are documented, half of these in the sea. Some of these fish lineages are ancient, like sharks which originated 400 million years ago. Fish live in nearly all marine and freshwater settings from the deepest or hadal regions of the oceans, 8 kilometres deep, to lakes at over 3 kilometres altitude. They have evolved a multitude of different ecological roles as predators, prey, filter-feeders, cannibals, cleaners, and parasites. But even this enormous range of habits, biomass, and biodiversity cannot insulate them from the voracious appetites of humans.

Our desire for fish has even outstripped our wish for meat. While human numbers grew by an average of 1.6 per cent per year from 1961, fish consumption grew at 3.2 per cent per year. Even though much of our fish consumption is now supplied by aquaculture, many wild fish stocks are being driven below sustainable levels. These fish are captured by 4.6 million fishing boats on the seas. Many of these are small vessels used by artisan fisherman working close to the shore and providing for their families and local produce. At the opposite end of the spectrum are the leviathans, weighing more than 100 tonnes, and there are about 40,000 of these. The largest of all is the *Damanzaihao*—now renamed the *Vladivostok 2000*, that weighs nearly 50,000 tonnes. It is twice the length of an American football pitch and can process over half a million tonnes of fish a year.

The human impact of overfishing has increased dramatically in the last few decades. In the mid-1970s just 10 per cent of global fish stocks were considered overfished, but by 2016 that had risen to one-third. In some areas of the sea with densely populated hinterlands this is approaching two-thirds of fish stocks overfished. And as the surface fish diminish, the nets seek ever deeper supplies. Amongst the more environmentally damaging of fishing techniques is seabed trawling, which involves dragging a net over the seabed to scoop up fish and shellfish. It is the equivalent of driving an excavator through a pristine woodland to collect blackberries (though the damage, in terms of smothering blankets of stirred-up mud on the seabed, extends more widely). Deep-sea trawling removes some 19 million tonnes of fish and shellfish each year,[29] but its damage to seabed communities with their delicate sponges, corals, and sea lilies is incalculable.

Whole groups of fish have shown rapid declines in their numbers, none more so than sharks. Our vivid fear of their predatory reputation disguises the fact that we are the existential threat to their continued existence. Over the past 50 years the oceans have lost more than 70 per cent of their sharks and rays, with many species now threatened with extinction.[30] Huge numbers—tens to hundreds of millions—of these fish were being taken from the oceans in the early twenty-first century, many simply for their fins, before overfishing cut the numbers available.

One illustration of this decline is the whale shark. It can live for over 100 years, growing to be the biggest fish in the sea, and one of the biggest that has ever lived. It can be over 18 metres long, a third longer than a London bus, but it doesn't fit the usual caricature of ruthless predator. This gentle giant ranges across great areas of the tropical seas, feeding by filtering enormous quantities of plankton from the sea. Occasionally, the whale shark is known to congregate in shallow waters, and one such place is Ningaloo Reef in Western Australia. Since the 1990s, information on the size and sex of these coastal visitors has been collected, and over that time the numbers of larger specimens decreased; they are now 2 metres shorter on average, and there has been a two-fifths reduction in their overall numbers.[31] Although the hunting of whale sharks is banned in Australian waters, whale sharks navigate over hundreds of kilometres, and the likeliest reason for their shrinking size is their capture for food and even more wastefully for their fins, by illegal fishing in the Indian Ocean and the seaways of South East Asia.

The whale shark is an easy standing target for illegal fishing, gentle, large, and obvious in surface waters. But even those fish hidden deep below the surface are now targeted, detectable by sophisticated radar, and susceptible to dragging nets; they too have nowhere to hide.

The slimehead is one of these fish. It lives in the perpetually dark and chilly waters of the deeper Atlantic, Indian, and Pacific Oceans to depths of nearly 2 kilometres. Often it will congregate in large shoals around submarine seamounts. It is a sizeable fish, over half a metre long and as heavy as 7 kg. It is not—to the human eye—very attractive, looking somewhat like a hangover from the Palaeozoic seas, with its large head and prominent jaw, and a network of mucus-bearing sensory glands on its head that gave it its common name. Looks can be deceptive though, and its sweet taste and silky flesh have made this fish a culinary delight. Renamed the orange roughy, recipes now abound for it with lemon and orange, or cooked in garlic butter, or baked with parmesan and dill (somehow, 'slimehead with Pico de Gallo' does not sound quite so palatable).

The poor slimeheads are recent victims of their tastiness to humans. They only began to be captured in the late 1970s as traditional sources of wild fish were depleted. Because the fish are sluggish, and form large shoals, they make easy pickings for bottom trawlers, with catches of up to 50 tonnes per minute being reported.[32] It is no surprise therefore, that after a few years, many tens of thousands of tonnes of orange roughy were being removed from the waters of south-west Africa, New Zealand, and Australia, drastically reducing stocks to well beyond what was sustainable. Some observers referred to an orange roughy 'gold rush', made worse by the long maturation of this fish—which lives for over one hundred years—meaning that fisheries would recover only slowly.[33] Bottom trawling has also damaged the seabed communities that support the orange roughy, leaving a vicious cycle of fewer and fewer fish.

As stocks of wild fish become further threatened, and ever more remote places are targeted for fishing, might one solution to our depleted oceans be farming?

The taming of the carp

The common carp is a freshwater fish native to western and eastern Asia and has been a staple of Chinese cooking since prehistory. This was interrupted—briefly—during the time of the Tang Dynasty (618–906 CE) when the emperor Li forbade its capture, selling, or killing, because its common name in Chinese was similar to his.[34] The common carp's domestication had already happened several millennia prior to the emperor's intervention, and probably began with people noticing and exploiting the natural spawning patterns of the fish, just as indigenous fishing people do today. People were cultivating the common carp in prehistoric China 8,000 years ago,[35] selecting individuals of a similar size for food and also building ditches to help control its presence. People then went one step further, deliberately building ponds to nurture the fish and perhaps even using rice paddies to breed and tame them. Much later, and accelerated by Emperor Li's decree, several other species of carp were cultivated. Carp farming became a sustainable practice, of benefit to the plants from the natural fertilizer produced by the fish, whilst the fish ate the insects attracted to the paddy.[36]

For much of prehistory and history the kind of aquaculture practised by the Chinese, and by many other peoples from Europe to Australia, provided a relatively small, but generally sustainable resource of fish, mollusc, and arthropod protein to humanity. Aquaculture was certainly practised in the Mediterranean from early times, growing out of the abundance of migratory fish, like the gilthead bream, that entered coastal lagoons like that of Bardawil on the Sinai Peninsula.[37] The Romans wrote of the abundance of fish in the Adriatic and constructed *vivaria* along the rocky shoreline to nurture them. Sometimes these were simple constructs of pools bounded by stones, and other times collections of rectangular basins that were carved into the local rocks, where fish could be bred and farmed. Some households were wealthy enough to own their own *vivarium*, a luxury accessory or status symbol for a Roman. Fish would be salted, fermented to make garum, or sometimes transported alive on ships that had their own 'in-built' freshwater *vivaria*.[38]

By the Middle Ages of Europe, patterns were emerging that would change humanity's relationships with the sea, as fishermen sought out ever more remote stocks of sea fish, like herring, in the waters of the Baltic and western approaches.[39] As wild fish stocks dwindled, aquaculture was employed to supplement supplies and thus began a trend, accelerating in the later twentieth century, where fish farming sought to fill the gaps in supply. This change became dramatic over the three decades from 1990, rising at an accelerating rate, from providing less than 20 per cent to half of our fish.[40] This intensification of fish farming has changed the ecological landscape of large parts of the coastal zone, with the loss of many natural ecologies like mangroves, the spread of viruses through coastal fisheries and antibiotic resistance, and then the spread of those antibiotics through the food chain to humans.

There is perhaps another way to farm the seas more successfully. One that would be closer to the more sustainable origins of aquaculture, one that requires the taming of many fish. Animals must sacrifice their freedom to be domesticated, acquiescing to enclosed spaces, accepting our overlordship and food, breeding in captivity, and having a good disposition to us— though the auroch took some convincing of this! They must also reproduce regularly and grow quickly.[41] These obstacles to domestication have been insurmountable for most land animals, and hence there are no herds of domesticated elephants or rhinoceros. Domesticating a fish might then seem unthinkable. After all, in nature they roam freely in a medium that is three-dimensional, difficult to enclose, and that covers over 70 per cent of the surface of the Earth. But here is the key to a potentially more sustainable aquaculture. For although only a few fish have so far been successfully domesticated, mainly salmon, trout, and carp—many others, indeed hundreds, exhibit patterns of behaviour that might enable them to be tamed.[42] And so, just as with the carp in ancient China, there is the possibility of rearing local species that harmonize better to their native ecologies, and if this includes herbivorous fish, one might avoid the use of feedstuffs like fishmeal or so called 'trash' fish.[43] That way we might be able to avoid some of the pitfalls of intensive agriculture on the land, with its few but hyper-abundant species, and preserve much more of the oceans' natural ecology.

Biting less hard?

Our collective appetite—just for food, discounting, for now, other needs and desires—is transforming the Earth from the paradise of biodiversity it used to be, into a biologically impoverished food lot. Collateral damage includes climate change (of which food production is a major driver), depletion of freshwater, degradation of soils, and pollution of rivers, lakes, and coastal seas, not least by artificial fertilizers washed off the land. These pressures can only increase as human population rises by a few more billion this century, as is projected—a population of which a significant part *even now* is chronically malnourished, even while an obesity crisis intensifies side by side with this hunger. Can anything be done to alleviate these pressures, while preventing an even greater proportion of humanity slipping into food insecurity?

If humans were collectively rational, then some of the steps would be obvious and simple. Eating less meat is the most obvious example, so that we take nutrition from plants directly, instead of second-hand from dead animals, often killed with great cruelty, and with huge energy losses incurred along the way. This would both greatly ease pressures on ecosystems and biodiversity and (with a little care in our diets) provide us with healthier food too. Veganism and vegetarianism are on the increase—but so far their effects are globally outweighed by the seemingly inexorable rise in meat consumption elsewhere: the amount of meat each average human consumed on Earth almost doubled between 1961 and 2013, for instance,[44] while the world's population more than doubled in that time too. Strong cultural forces are at work here, albeit variously expressed in different parts of the world, and the culture wars can flare particularly fiercely where food is concerned. There are intense commercial pressures, too, as the food industry is a huge one, where profits and livelihoods are jealously guarded. So, amid these swirling and powerful forces, what kinds of things might make a difference?

Some things might. The Earth's food landscape is poised to see major changes, and how these will play out will be a major factor in determining how much of Earth's biosphere will survive over coming decades and centuries, and in what shape it will be. Some trends are already advanced, and poised for further change. Technology seems destined to become even more widespread, as we run out of fertile wilderness to develop into farmland, and need to extract ever more from the land that we have got. This need not necessarily mean an inevitable transition into ever more gargantuan plots of land in submission to mechanized monoculture, with blanket inputs of fertilizers and pesticides, and the whip hand being held by multinational corporations, while small farmers are forced out of business and/or into subservience (though such dangers clearly exist). Farming practice as transformed by robotics, sensors, and drones might have a much more nuanced—and perhaps in some respects a lighter—footprint on the land, and perhaps on people too. Drones equipped with sensors can already survey a field to search for small incipient patches of infestation, and then target just those to spray, rather than blanket-covering the whole field with pesticides. And, those herbicides and pesticides are needed largely *because* of the monocultures, which are an open invitation to biological attack. Light, wheeled robots with sensors could map the complex distribution of soil types, and use those maps to plant (and harvest) complex mosaics of different crops side by side, as best suits these intensely local patterns of soil moisture, nutrients, and topography. These kinds of developments—which are already technical reality[45]—would place the local knowledge of the small farmer at a premium, and perhaps help tip the balance against today's monolithic 'advanced' agribusiness methods.

There may be a yet more fundamental nut to crack. The underpinning of all of agriculture—and indeed of most life on Earth—is the energy of sunlight captured through photosynthesis. One might have thought that, following nearly 3 billion years of evolution, this would be a honed and super-efficient mechanism. Not so. Photosynthesis is terribly inefficient, with only about 6 per cent—at most—of the sunlight falling onto a leaf being converted into biomass. Why so? In the natural world, there are so many powerful constraints on plants—limited nutrients, competition, pathogens, climate—that simply bulking up leaf mass more efficiently does not necessarily come high up the list of priorities. And, the complex reactions involved in photosynthesis evolved on the ancient, and very different, Earth of the Hadean and Archean eons. For instance, one of the major pathways in photosynthesis is controlled by an enzyme called rubisco (it is probably the most common enzyme on Earth), which controls the fixing of carbon dioxide into sugars. But, it does not distinguish oxygen from carbon dioxide very well, and often fixes it instead to make a toxic compound called glycolate—which the plants then have to work hard to remove. Back in the Archean, when there was little oxygen on Earth, that was not so much of a problem, but by the time oxygen levels rose in the atmosphere, this complex photosynthesis mechanism together with its built-in error was thoroughly entrenched. Plants subsequently have simply lived with it—at the cost of using about a third of their energy resources in getting rid of the glycolate. Researchers, using both 'traditional' means of plant breeding and genetic engineering, are now wrestling with these fundamental brakes on plant productivity.[46] Some approaches are looking to modify the rubisco mechanism, while others seek to develop pathways through which plants can cope with glycolate more effectively, and progress seems to have been made, at least in the laboratory and on the experimental plot.

One research direction here is ambitious: to re-engineer rice, one of the world's major crops, to be a fundamentally different type of plant. One of the major evolutionary changes that has taken place in the geological history of photosynthesis is the emergence of what are termed C4 plants (in fact they have evolved independently more than 60 times). This form of energy conversion involves a four-carbon molecule being produced as a step in the process, instead of a three-carbon molecule as is typical of 'normal' C3 plants. C4 plants, such as most grasses, can be up to 50 per cent more efficient at capturing sunlight, especially in hot and dry conditions. Rice is a C3 plant, but for more than two decades scientists have been looking to supercharge it by converting it to be a C4 plant. The remodelling would restructure the plant at all levels, from its basic anatomy to its molecular machinery, and would need all of the modern alchemy of genetic engineering—and even so is said to be at least two decades away yet.[47] Nevertheless, this dream, being actively pursued, is a sign of change to come. Only time will tell whether its promise will be fulfilled—and what, if any, hidden nightmares it might have in tow.

Among these developments are those aimed at detaching the production of meat from the raising and killing of animals. The idea has been there for some time. In 1931, Winston Churchill, then an up-and-coming politician, wrote an essay called '50 Years Hence'. In it he said, *'We shall escape the absurdity of growing a whole chicken in order to eat the breast or wing, by growing these parts separately under a suitable medium.'* Well, that absurdity is still with us, grown to monstrous proportions, and it has taken nearly a century to overcome the enormous difficulties of reproducing meat in the laboratory. But the first serious attempts are now with us, and they take a variety of forms. One way is to cheat, and disguise plant material well enough to make it taste like meat, especially within that seemingly inescapable burger format. There is the Impossible Burger, for instance (a concoction of wheat, potato, coconut oil, and yeast—the last of these yielding the haem protein that otherwise comes from blood), and the Beyond Burger (with beetroot juice adding the red tinge). They're nutritious, not so far off the real thing in taste, and have a much smaller ecological footprint than the real thing. They're not quite mainstream yet—but have already sparked enough of a backlash (for not being 'natural' enough) to suggest that this trend might grow. One can partly cheat, as in insects such as crickets and mealworms now being considered as a potential source of animal protein, Or, one can grow the real thing, by harvesting some stem meat cells from an animal, culturing them in a bioreactor, and converting them into meat. That's more of a challenge—but is also now technically possible, and is being done. It's tricky, as a three-dimensional 'scaffold' is needed to grow the cells into tissues, and fat as well as muscle cells are needed for a lifelike product. And so far it has also been very expensive (at thousands of dollars per kilogram). But, the start-up companies are already in place, and searching for the breakthroughs (and the funding) that can make this process commercial on a wide scale.[48] They hope that once these new burgers are cheap enough and tasty enough, they will sweep the world, in the way that computers and mobile phones have done within a few decades.

Could that, then, at last play a major part in relieving the pressure on the world's remaining wild places? Time will tell—and one suspects that, for all of the ingenuity of the technology, and the speed at which it is now evolving, it is the human factor that will be the critical one. Nevertheless, the technological factor is now enormous. For our planet it really is something new under the Sun, and in some ways it has become as large a player as the biosphere. We tell its story next.

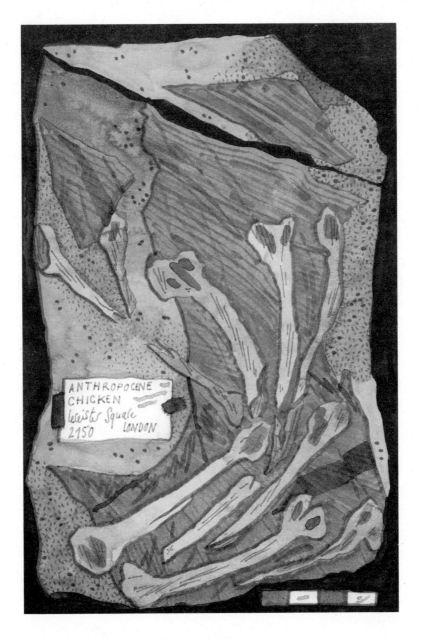

7

MIRROR TO THE WORLD

On the early Saturday evening of 23 November 1963, electromagnetic waves were beamed out from tall metal masts across Britain. They carried within them faithful replicas of human movements and words that had, a little earlier, been captured as a combination of microscopic silver crystals and magnetically aligned iron particles, ingeniously ingrained into long loops of plastic film. These patterns were then converted back into patterns of light and sound via cleverly manipulated cathode rays and electrical currents in heavy square boxes that were, at exactly the same time throughout the land, brought to life by 4.1 million other humans, as all these humans had been cleverly manipulated in turn, to absorb a dream for the next half hour. The dream (repeated religiously each week thereafter) was of yet more astounding marvels, amid high adventure, to arise in a promised future. The first ever episode of *Doctor Who* was about to be aired, to change the mental landscapes of its human recipients.

The scatty and irascible old Doctor was an unlikely hero, the Tardis of banal police-box exterior was the quirkiest of spaceships, and his nemesis, the slowly trundling Daleks, were villains in the most absurdly pop art style—though their merciless extermination of all non-Dalek life-forms was watched, by many duly terrified adolescent humans, from that safe space behind the sofa. Nevertheless, this inspired piece of science fiction gave at least some of its viewers their first inkling of the wonders of the cosmos, and of the near-miraculous potential of technology.

That technology was in full swing by the time *Doctor Who* was screened, forming an indispensable backcloth to the lives of the viewers. The emerging television network was just one more obviously visible part, while the space race developing between America and Russia produced real-life derring-do that reached out to the cosmos. But many more prosaic things were happening, too, that in their own way were just as extraordinary as the feats of the *Apollo* and *Vostok* spacecraft—and that had a much deeper impact on Earth.

The day after the *Doctor Who* ritual, the young devotees would have time free to play—but the following day they would, reluctantly, be wrapped in woven fabric to protect them from the cold and put into wheeled transporting machines to be taken, along hardened strips of ground that snaked through the countryside, to a gathering point. There, organized into groups within large, baked-clay constructions, their elders would, day by interminable day and year by never-ending year, teach them something of how their world was built—and how they, in turn, in years to come, would maintain and extend that world. This was their empire. Their grip on it would grow forever tighter, so their lives would be ever more comfortable, even while their perspectives stretched out to the stars.

This world—including the make-believe world of *Doctor Who*—is replete with things that have been *made*, to help ensure survival, or just make life easier, or more amusing. These artefacts have become so ubiquitous and so ingrained in our lives, now, that we rarely notice their presence—though we keenly feel their absence if for some reason they are taken away from us. They are all things that, whether simple, like knives and forks, or complex, like digital cameras and television sets, have been puzzled out by the human mind and then physically constructed to serve some kind of purpose. They are the fruits of human technology, based on materials that extend back deep into geological history. Our technological creations, together with us, seem to be creating a new planetary presence, that is now evolving at high speed.

The process, though, had a very slow start.

The history epoch

The stone tools used by early humans predate our own species, *Homo sapiens,* by more than two million years: rough choppers, axes, scrapers, they show only glacially slow change from crudely broken rocks to somewhat better-shaped implements over this time. Generation after generation, our human ancestors copied their elders; innovation was on the menu only very rarely. A new step took place some 300,000 years ago, when the systematic fracture of 'flint cores' allowed a greater number of useful fragments from one starting block: a very early form of production line. Then, about 50,000 years ago, came the discovery that pressing on a rock with a sharp point—with care and experience—could shape it, by 'pressure flaking', more precisely than just by hitting it. From about 30,000 years ago came a flood of tiny flakes, 'microliths', seemingly made to embed into wood to form spears and harpoons. Innovation was beginning to speed up.

Wander through the North African desert, and you can, here and there, stumble upon beautiful flint arrowheads—some leaf-like, others triangular, many with carefully made notches to fit into a wooden shaft. This newly sophisticated manufacture shows not just more beautiful handiwork, but patterns that evolved, now over just a few millennia, and not over hundreds of thousands of years. The changes mark the start of the Neolithic, or 'New Stone Age', that started more or less when the latest of the Pleistocene ice ages gave way to today's warmth.

Hunter-gathering—the lifestyle that had, previously, wholly sustained *Homo sapiens* and all other species of humans for millions of years—then began to be replaced by agriculture. Humans settled, to tend to their delicate, newly domesticated crops and obstreperous animals, and their numbers grew. For a few there was less physical work, if they were strong or cunning or ruthless enough to seize control, and have others work for them. Hierarchical societies grew. This early settled human life was probably more miserable, by and large, than the hunter-gathering one (there is an old joke among archaeologists that fermented grain, and thus beer, was invented as consolation for this new servitude). But, it provided more calories, and hence was more successful.

Within this 'agrarian' revolution many new technologies evolved and developed. Pottery, for example, which had been invented 10,000 years earlier in some Far Eastern cultures[1] was reinvented in the 'Pottery Neolithic' of the 'Fertile Crescent' of the Middle East about 10 millennia ago, and the idea and the skill to make it migrated, as word of the usefulness of this new invention for cooking and storing food spread. Weapons became used less as instruments to kill other animals and more to kill other humans. And, someone more than five millennia ago—perhaps while working in a pottery kiln—saw a strange red-hot fluid puddling on the ground. When cooled and hardened, it formed a hard material that could take a wickedly sharp edge. This was metal, and new and more deadly sorts of weapon were born from this new material. It could be shaped into use for protection *against* weapons, too, to fuel a race between arms and armour that remains of deadly intensity today.

There came the invention of things like the wheel (a tricky thing to construct initially, from wood), money, woven fabrics for clothes, the saddle and bridle, the plough, scythe, and spade, writing, new kinds of buildings, aqueducts, lamps, roads, bridges, and (inevitably) new kinds of weapons. It is the material story of the Holocene epoch, a flowering of the creation and production of artefacts, the remains of which now form a diverse and varied terrain for archaeological study. It is a new kind of evolution—albeit being channelled through exactly the same biological species, *Homo sapiens*, that had managed such slow and limited technological progress for so many millennia previously. Indeed, this early halting progress was not much more sophisticated than that achieved by its fellow human species, *Homo neanderthalensis*, and one wonders whether the Neanderthals might ultimately have been the vehicle for a similar technological lift-off, if they and not our own species had survived.

Nevertheless, for all of this new material sophistication, there were limits to the spread of technology—and of humans. The energy that people could obtain to do work—in building, farming, warfare—was ultimately limited to human muscle power, to the muscle power of the animals that humans forcibly co-opted, and to inventive but limited attempts to harness the energy from streams and rivers, in watermills and weirs, and the wind, in windmills and sailing ships. The technology could be ingenious, but it was mostly artisanal; the continuous input of physical and mental effort and learned skill set a limit on the amount of manufacture. Humans were limited nutritionally, too, by the crops they could grow and the animals they could rear—and these organisms were limited in turn by the nutrients in the soil, notably nitrogen and phosphorus. Even with fewer than a billion people on the planet, hunger and starvation were always around the corner. And, for those that managed to stay well fed for a while, there was always some kind of plague or pestilence to threaten them. Those people travelled, and migrated, to trade or to try to reach more promising lands, but that travel was long, arduous, and dangerous—an adventure that all too easily and often finished in tragedy. The world of Holocene civilization was connected, but only semi-connected. Many people did not leave their towns and villages in their lifetimes, and different parts of the world developed their own distinctive cultures, with influences from trading, travel, and imperial conquest modifying, but generally not overwhelming, these differences.

One might call it a sustainable world, for all the hardships and dangers that its human (and non-human) inhabitants endured. The basic support system for all of this life was in place and more or less stable, at least as far as memories and ancestral stories went: soils and the sustenance they provided; clean water (though this was increasingly in question, where humans gathered in larger numbers); the 'immemorial' pattern of land and sea, upon which property and empires alike were based; and the seasons followed each other, some better and some worse, but with the sense that this pattern, too, was simply woven into the fabric of a capricious but fundamentally everlasting Nature. Into this world, civilizations rose and fell over the millennia, but neither the well-being of humanity, nor the natural world, was seriously threatened.

But something was about to change.

Breaking the limits of nature

Wood has been humanity's main fire-provider for most of our long species' history (and was fire-provider to ancestral human species too). It continues to be used in large amounts, whether in rural Africa, South America, and Scandinavia, or as imported wood pellets in power stations in the UK. But the fires can only be kept burning for as long as there is wood—and the trees that supply it, once cut down, can only grow so fast. The need for fire, for heating, cooking, and smelting metal, led, as the millennia of the Holocene passed, to forests progressively being stripped from landscapes across the world. For more of this plant-sourced energy, one had to wait until the trees grew back again. It was a natural brake upon human growth.

The discovery of coal, the rock that burns, added another source of fire—moreover, of a fire that burned hotter than wood, and that gave out more heat, kilo for kilo. It has probably been used, occasionally and serendipitously, for many thousands of years. Its systematic use, though, seems to stretch back some five millennia, to northern China, where trees are scarce and coal seams lie close to the ground surface. The ancient Greeks used it, as did the Aztecs, and the Romans who, when they colonized Britain, found and won coal from most of the major coalfields. But to do any more than dig out any coal that happens to lie at the surface is a brutally difficult and dangerous business, as any coalminer knows, with the danger of rockfall, underground flooding, and explosion ever present. So, across those millennia and until little more than a couple of centuries ago, the efforts to extract this subterranean energy were—as far as the Earth was concerned—little more than scratching the surface. Such artisanal extraction and burning of coal had a negligible effect, for instance, on atmospheric carbon dioxide levels.

That began to change a little more than a couple of centuries ago, as technology and energy started a new phase in their relationship, to solve a problem that had bedevilled human use of coal—and, indeed, the extraction of anything from deep underground. Dig a deep (sometimes not all that deep) hole in the ground, and it will fill with water, because rocks at depth are soaked in water that has fallen as rain on the landscape and then gradually percolated down. Go below the level at which this subterranean water collects, the water table, and you have to be permanently baling out the water if you want to keep a mine or quarry safely dry. Coal mines could only go with great difficulty beyond shallow depths, and mine owners searched constantly for ways to remove the water that stubbornly kept flooding in. Another form of water—steam—gave them the answer.

Heat water until it boils, and the steam is not only scaldingly hot, but exerts a strong pressure as it expands. This has long sparked curiosity. The Greek mathematician Hero (or Heron) of Alexander (about 10 to 70 CE) built a curious machine called an aeolopile—a water-filled cylinder on an axis with a couple of angled nozzles coming out of the sides. Heating this contraption made steam gush out of the nozzles, making the cylinder rotate. For Hero, it was probably mainly a party trick, but a generation earlier, the Roman architect Vitruvius (about 75 to 15 BCE) commented of an earlier protype that the 'violent wind' that it produced was testament to some 'divine truth lurking in the heavens'.

The seventeenth and eighteenth century mine-owners were probably little interested in divine truth, but when the English inventor Thomas Savery proposed using the power of steam to force water out of coalmines and metal mines, they pricked up their ears. The early designs were crude, only partially successful—and dangerous too, being liable to be explosively shattered by the high-pressure steam. But improvements to the design by first Thomas Newcomen (1664–1729) and then James Watt (1736–1819) made them more effective, and soon ubiquitous in deep mines. These early steam engines needed a lot of energy—from the burning of coal, of course—to make them work, but the release of that energy allowed hugely greater amounts of coal to be hewn from ever deeper within the Earth. This yielded energy that was very soon expended in other forms of steam engine, put to use to make factory machinery more powerful and productive, and to power locomotives and ships. All of this hugely increased the demand for coal, and allowed the development of yet more different kinds of machine by those ingenious and indefatigable Victorian-era scientist-engineers. Humanity began to lose its dependence on the normal resources of Earth which had sustained it—as it had sustained so many other species—for so many millennia, and began to explore a lifestyle that, until very recently, has seemed to be almost limitless in its use of natural resources.

A few decades later, oil and gas joined coal as energy providers, and they also spawned families of new machinery designed simply for their extraction from their hiding places deep underground: drilling rigs, supertankers, refineries, oil platforms. And these new energy sources, fluid and supremely manageable, generated a further cornucopia of machines: automobiles, planes, earth- and rock-moving machines, metro systems, and much more. A fire had been lit that would soon be an explosion—a *constructive* explosion, out of which millions of new species bloomed. These were not species of flesh and blood (or at least mostly, they were not). They were technological species, 'technospecies'. And they would soon, in some important respects, take over the world.

We are now surrounded by so many things, such an enormous variety of *stuff*, that we take this for granted. We are a supremely adaptable species, and this new material hyper-diversity has become part of the background of our lives, to the extent that we can barely imagine life without the almost limitless opportunities it provides. How could we cope, now, without computers and mobile phones, without ballpoint pens, lamps, knives, forks, clocks, chairs, combs, scissors, refrigerators, ovens, TV...?

And what are these things, these species of modern technological invention, really? We generally call them all artificial, or artefacts—that is, things not found in nature—and so in our minds they are mostly separated off from nature into a new category that has nothing to do with biology, the biosphere, or ecosystems. And yet *we* are biological organisms. We have developed the capacity to make tools, of course—but we have already seen that other organisms, like the New Caledonia crow, can fashion and use tools—and that humans genetically identical to us used tools not much more sophisticated than those devised by the New Caledonia crow for most of the timespan of our species. But, there is a profound difference in quality between a Stone Age hand axe and crow-fashioned twig and, say, a television set or lightbulb. There is also a profound difference in *quantity* between a hand axe/shaped twig and all of the manufactured objects that surround us, when we are comfortably sitting and watching television. And, yet more, there is a difference in *connectedness* between those individual ancient simple tools, and the complex material networks that allow, for example, an episode of *Doctor Who* to be made and screened around the planet. So, what has happened—indeed, *is happening*—to cause the emergence of this new kind of world?

Counting technospecies

One of the problems in trying to understand this new world of technologically made objects is its sheer profusion. How many kinds of human-made objects are there out there? And how rapidly are they multiplying? As far as we know, no one is even trying to make any overall census of this (not least as it is now a permanently, and rapidly, moving target), but we can at least try to set the phenomenon of technological species into some kind of biological perspective (where some numbers at least do exist, for comparison). We can link our technological artefacts back to biologically made or shaped objects or structures made by earlier human species, and also by non-human species, and here we can move into the realms of biology and, perhaps more surprisingly, palaeontology, as we seek this perspective.

We might start by trying to estimate the number of biological species present on Earth today, which includes the number of named and catalogued species alive, and those yet to be discovered and named. Even the former number, a little surprisingly, has various estimates, but they centre around about 1.5 million. Estimates of the number of species still to be found and described have ranged from about three million to over a hundred million, more recent ones being of the order of 10 million species.[2]

From palaeontology (which recognizes species in a different way to biology, but not so differently that inter-comparison cannot be made) we know that the average species known from the fossil record exists for something between one half and five million years. If we take 3 million years as an average, the last half billion years of abundant multicellular organisms should have seen the appearance and then extinction of about 1.5 billion species. Now, the number of species discovered and catalogued by palaeontologists (who work slowly and carefully, it must be admitted) is around a quarter of a million—and so only a tiny fraction of ancient biodiversity has so far been recognized. This is because most species are soft-bodied, and so fossilize only very rarely, if at all—but also because there are many fossil species still to find out there in the rock strata (between us, in our careers we have found and named a few dozen species of fossils, and if we but had time to do more active palaeontology, we would undoubtedly find more).

There is a category of biological phenomena, and of the fossil species that can form from them, that is a little special, and that takes us a little closer to the realms of technology. These are the traces that animals leave behind as they move, a category which grades into and includes the constructions that some of them build. These are things like footprints, animal burrows, and insect nests. When fossilized in strata, they are called 'trace fossils' by palaeontologists, to distinguish them from the 'body fossils' represented by things such as bones and shells. In non-human biology, one of the features of these traces is that one animal tends to make only one or a few kinds of traces, each being more or less directly from specific coding for this purpose in their DNA. Thus, each species of ant, termite, or wasp, tends to build the same kind of nest, albeit often adapted around the local small-scale topography of the ground. Each spider species has its own pattern of web, and each burrowing organism has its own pattern of burrow. Each animal can have a range of behaviour, of course: the iconic fossil trilobites of the Palaeozoic Era, for instance, are well known for producing three types of trace, depending on what they are doing at any particular time: a set of V-shaped marks called *Cruziana* from ploughing through the sediment; an oval depression called *Rusophycus* from settling into a little scooped-out depression on the sea floor; and 'tip-toe' walking marks called *Diplichnites*. And of course, some very different animals can make the same kind of trace. While *Rusophycus* is most usually associated with trilobites, very similarly shaped depressions can be formed by particular kinds of worms, snails, and shrimps.

The lesson here is clear. The ways in which animals reshape their local environment, to the extent of making constructions—and which extends to occasionally using tools—is limited, and closely coded by genetic make-up. We are not aware of any total counts of named trace fossil species, but probably those recognized so far—for the whole of the geological record—might amount to several thousand.

Now, the various ways that we reshape solid material around us can be considered as a form of trace, whether it is a footprint (where the relation is clear), or a building (which is directly comparable to a wasp or termite nest), from which the analogy can be extended to the electrical fittings, the carpet, television set, the cutlery, and furniture—and the car parked outside on the driveway, and mobile phone in the pocket of the driver.

The difference, of course, is that we are but one species, and what we have very suddenly developed is the ability to conjure up a range of artefacts of such near-infinite variety that making an estimate of 'technospecies numbers' is an even more daunting task than trying to measure biological species numbers on Earth today. Examples of technospecies may include a Swan lightbulb from 1880, or the Osram version from the 1920s. Or a Bic Cristal ballpoint pen, which is instantly recognizable—and abundant too, well over 100 billion having been manufactured since its commercial release in 1950. There are many other kinds—and so technospecies—of ballpoint pen, each sharing general characteristics of ball, barrel, ink-filled cylinder, and cap, but recognizably different in detail. These might all be regarded as being members of a genus of ballpoint pen, in a similar way that our species *sapiens* is just one within the genus *Homo*, along with the species *habilis, erectus, neanderthalensis*, and so on. And what of pencils, fibre-tip pens, fountain pens, crayons? These might be regarded as representing other genera within a family of writing implements, each with their own myriad component species, a little like we belong to the family Hominidae, that is the great apes, along with gorillas, chimpanzees, and orangutans. It is important not to push this analogy (and it is only an analogy) too far, because these relationships in technospecies are based on similar morphology and function, and not on evolutionary connections acquired over millions of years. But equally, the analogy is important to try to gain some appreciation of this phenomenon that is unfolding around us among these so-newly evolving genera and species of technospecies.

The ballpoint pen genus is less than a century old, invented by the Jewish–Hungarian journalist László Bíró in the 1930s.[3] Noticing that newsprint ink dried more quickly than fountain-pen ink, but was too thick to flow, Bíró came up with the idea of placing a small rotating ball at the end of a tube of this viscous ink. From this he obtained fame, but not fortune. In those turbulent times of danger and racial persecution, he had to sell all his shares in the fledgling company to allow him and his family to flee to safety (he did not regret this bargain, made of necessity).

The fountain pen that he helped to relegate to obscurity has a longer history,[4] though mostly of rare, isolated attempts to make a writing implement that worked consistently and without too much mess. The Ma'ād al-Mu'izz, the caliph of the Maghreb, in north-west Africa, wrote in 973 CE of having an ink-filled tube made for him, and in 1663 the diarist Samuel Pepys mentioned talk of a similar contraption—but functional and widely available fountain pens only really appeared from the mid-nineteenth century. The fibre-tip pen is younger even than the ballpoint pen, first appearing in 1962. Even the humble pencil is just a little more venerable by comparison, the trick of placing a graphite stick in a small wooden casing only being worked out in 1795, by Nicholas-Jacques Conté,[5] an officer in Napoleon Bonaparte's army. A 'universal man', according to Napoleon, he invented the crayon, for use by artists too, for good measure. This new plethora of writing instruments overwhelmed the basic writing machinery that had been used for many centuries—a bird's feather or reed cut diagonally to make a nib, or a piece of chalk to write on a slate tablet—and, incidentally, brought writing within the range of the whole human population, and not just a few select scribes.

How many individual technospecies of pen and pencil have been designed and manufactured since then, in innumerable factories around the world? Just rummaging in a single office or study will normally turn up several dozen. We would guess tens of thousands, but it may well be more. The same kind of uncounted, perhaps uncountable, profligacy holds true for so many of those familiar household objects that populate the world that we now live in. Technospecies counts are rare. One example is a count of the published book titles that now lie within Google's vast electronic databases. Each can be thought to represent a single technospecies, having a specific pattern of page numbers and shapes, printed word patterns, covers, binding. Some, the bestsellers, are abundant, produced in their millions; others are rare, some slim volumes of delicate poetry being lucky to see a print run of a hundred—but such a range of abundance holds true for biological species, too. The grand total of book technospecies, totted up by some exceedingly patient Google employee in 2010, was 129,864,880. And, each year some 2 million new book titles are added to this mountain of narrative (the book you are reading is one of them), and so by now the number has likely risen to about 150 million. Extraordinarily, the technodiversity just within the book family (admittedly a large and extended one) exceeds modern biodiversity more than tenfold. Add in all the other technospecies and it may be—who knows?—perhaps a hundred-fold. It is a kind of evolution which, for so much of human history a thin trickle, has turned into an explosion.[6]

We are living amid the full power of this continuing explosion—what is it doing now? Commonplace items, like pens and books, help to give an idea of its scale, but one might say that these are almost infinite variations on relatively simple themes. A spin-off from this technodiversity explosion gives another, and perhaps deeper perspective: the synthesis of new kinds of solid inorganic chemical compounds in the laboratories of the materials scientists. In nature, these are known as minerals, and the Earth has, over its history of 4.5 billion years, produced a little over 5,000 different kinds, as patiently tabulated by mineralogists. Some are familiar and common, like quartz and calcite, while many are exceedingly rare. Humans began to make new minerals, not present in the natural world, when they began to do things like extract metals from rocks and make new alloys. But their capacity to do this long remained limited—until new technical possibilities came, in the form of laboratories with ever more sophisticated equipment. There is a database that shows exactly how those possibilities flowered, in the German city of Karlsruhe,[7] and for years it has been patiently collecting data on all the new kinds of mineral that humans have suddenly developed the capacity to make—and 'suddenly' is the exactly appropriate term. In the early part of the twentieth century a few hundred extra minerals were noted. By the 1930s it had reached about 1,500, and by 1950 this had roughly doubled to about 3,000. Then the boom began. By the early 1980s, 30,000 synthetic minerals had been synthesized. By 2000, this number had more than doubled to over 70,000. Now, another two decades on, the number of synthetic minerals has tripled from that to over 200,000. Currently about 7,000 new minerals—more than the Earth managed to produce in 4.5 billion years—are synthesized each year. It is symbolic of this sudden explosion of technological possibilities, and is an outburst of chemical creativity that may be unique in the cosmos.[8]

This extraordinary diversification is not driven by any change in the biology or genetics of the humans involved, for we are no different in this respect from our Stone Age ancestors. It is driven by patterns of circumstance and history, and on the back of generations of ever more sophisticated technology devised, ever more rapidly, by our ancestors, mostly our very recent ancestors. Indeed now, that 'ancestor' is often a younger version of the same human individual, as major technological advances are made from one decade to the next. We see the most obvious manifestations as they immediately affect us, as a stationary phone line is replaced by a mobile phone, and that in turn by (now, several generations of) smartphone. Behind that, though is the kind of global hinterland which can, say, increase the world's mineral content many times over, as just one globally interwoven system involving accelerating levels of knowledge, energy, materials, and finance. These are refashioning the Earth's surface materials at a speed that is now dizzying.

All these structures are biologically made, in the sense that we are a biological species, but something else is going on, and not only because our biological skills and aptitudes are now increasingly augmented by computers, with silicon-based computational capacities and speeds far outpacing human capabilities.

A new planet-spanning entity of technological structures has, very recently, arisen on Earth. Peter Haff of Duke University in North Carolina, an extraordinarily insightful combination of geologist, engineer, and philosopher, calls it the technosphere[9]: a new 'sphere' on the Earth's surface to join—and intersect with— the lithosphere of the solid Earth, the atmosphere, hydrosphere, cryosphere, and biosphere. Like the other, ancient 'spheres' of the Earth it is much greater than simply being the sum of its individual parts, and has 'emergent' properties and behaviour, many of which cannot be easily predicted—or controlled.

Weighing the technosphere

Something like the technosphere has been present ever since humans began making tools, but for most of this time it has been as small local entities, mostly thoroughly embedded within, and not greatly perturbing, the dominant biosphere. But, almost without us noticing it—mainly because we have been too busy building its many parts—it has coalesced and grown, very recently, into something that is gigantic and planet-spanning. The exploding diversity of most of its parts is difficult to measure, as we have seen, but its physical bulk seems to be a little easier to assess, albeit in different ways. One means is by using the data of the scientists who study the material flows of the economy—the raw materials we mostly take from the Earth, and the things we make out of them: buildings, roads, bridges, dams, railways, vehicles, and all else. At the beginning of the twentieth century the mass of all of these things that were in use was estimated to weigh in at about 3 per cent of the mass of all living things on Earth (that is biomass, measured as dry weight). By the mid-twentieth century, the working physical technosphere had grown to be equal to about 7 per cent of the biosphere's mass, but this growth was now accelerating, so by 2000 the value was about 50 per cent. It is still accelerating, and 2020 is the year when this active and functional technosphere has become equal in weight to the biosphere[10]— and is set to quickly pull away to grow yet larger. Over the last 120 years, the living biosphere has diminished by something like 5 per cent, mainly as forests have been cut back to make way for cities and agriculture;[11] the active technosphere has mushroomed by some 5,000 per cent, mostly in the last half-century. It is an astonishingly swift and dramatic entry onto the planetary stage.

The calculated mass, that is possessed equally by both active technosphere and biosphere, is a little under 1.2 trillion tonnes (i.e. 1,200 billion tonnes), of which a little under half is that modernity-symbolizing synthetic rock, concrete, in which metals such as steel and aluminium, and recently plastics too, figure as ingredients. This is a large amount—about two and a half kilos of manufactured, currently functional constructional material per square metre of the Earth, both land and sea.

But there is more. We are a terrifically wasteful species and have discarded far more than we use. So, to add to that figure, there are all the constructions that once served us a purpose, but are no longer in use—whether dismantled, bulldozed away, or simply discarded casually or to fill the growing landfill sites. There is also all the waste material generated when resources are extracted from the ground; this can vary from a modest proportion of the total (when sand and gravel are extracted from the ground, say) to vastly outweighing the resources themselves (in mining copper, typically more than 99 per cent of the material excavated is waste). There is also the material that we move in ploughing soil, and trawling sea floors, to keep alive the humans (almost all of us) who are involved in keeping the technosphere going. Factor all of that in, and the total amount of material that we use, or have used and discarded, is something of the order of 30 tril-lion tonnes.[12] About half of that is concentrated in and around the world's urban areas—but spread it evenly across the planet, and there would be about 50 kilos of material per every square metre (again, land and sea) of the Earth's surface. On such a world, thus, we are about ankle-deep in the things we have made and (largely) thrown away.

The human component

Peter Haff's concept of the technosphere is not just of some inani-mate robot-like growth of concrete and steel, copper, and silicon, that is outgrowing its ancient biological parent. There is—for now, at least—its human component, for it is human hands and minds that have built and shaped its myriad components, from toothpicks to skyscrapers. But the human role is not just to cre-ate and maintain it, but to be an integral part of this new sphere, caught up within it and, in all too many ways, trapped within it and dependent on it. For it is the global technosphere, and its ceaseless, ever-shifting transglobal flows of energy and mat-ter, that is keeping virtually all of us alive. Peter Haff notes that in a world where humans were simply yet another normal part of the biosphere, living by hunting and gathering—as we were for most of our species' existence—the Earth could support a population of just some few millions. It has been our ability to shape the environment around us that has allowed our numbers to grow, at first slowly and, much more recently, to skyrocket. Many individual steps have contributed to this. A key step, many thousands of years ago, was the development of agriculture, a way of life that could feed more humans in a smaller area than could hunter-gathering. But even after that, the human popu-lation grew only slowly, pegged back by the normal carrying capacity of the land. That barrier was shattered in the last century, by directed, hydrocarbons-fuelled, energy, that led to, among many other things, the energy-intensive manufacture of nitrogen fertilizer directly from thin air.

It is therefore a sphere that maintains humans in completely unprecedented numbers (for a primate) on Earth—but Haff argues that it is not as simple as that. The main function of the technosphere is to keep *itself* going, not us. We are completely tied into it and forced to maintain it, but have not directed its creation and development, as if we were a unified global society directed by some kind of omniscient and far-sighted world government. Rather, a combination of expanding technical possibilities and individual human discoveries have thrown up novel structures here and there, like spores landing on a petri dish. Those that in modern times landed on fertile ground—where there was infrastructure, investment, finance, a trained workforce, a marketplace—could take root, develop, multiply, and expand their territory, at extraordinary speed. Phenomena like the internet and smartphones are classic general examples, but there are many prosaic things too—CDs, e-cigarettes, selfie sticks—that, in a few years spread through human populations across the whole Earth.

What is more, these accelerating technological invasions have happened even as human society is (and seems to have become more) highly fractured, divided into competing—and often warring—nation-states, and with these internally divided into fiercely competitive political groups and industrial clans. In this state of affairs, there is little chance of unified control or direction. Regardless, the technosphere carries on growing and evolving, its planetary interconnections ramifying and diversifying, as new inventions dreamed up by the harried humans enmeshed within it are quickly taken up and incorporated into its ever-expanding repertoire of ways of transforming matter and energy.

Many of the functions and controls are increasingly automated, through the instantaneous electronic connections of the global internet that are now woven ever more tightly into it: the algorithms that run the speed-of-light transactions of modern stock exchanges, the databases for everything from banking to health to national security to science, or the programming of drones for modern automated warfare. Within this, artificial intelligence is still largely a facilitator of human-made plans, and we are (it seems) still some way away from the science-fiction scenario (or 'singularity', as it is sometimes called) where super-intelligent computers take over to run things for themselves. The technosphere still needs humans, as much as humans now need the technosphere. And humans, of course, are not the only biological components of the technosphere. To keep them—us, that is— alive in our present numbers requires huge nutritional inputs, supplied by the expanding agricultural landscapes that continue to grow at the expense of natural wilderness. To make these productive enough to feed us all requires the continuous expenditure of fossil-fuelled energy and inputs of high technology (one might recall the broiler chickens, pitifully short-lived, and helpless outside of their closely controlled enclosures): modern agriculture may still be technically part of the biosphere, but it is firmly enmeshed within the technosphere.

But the situation is not stable, is moving fast, and may be prone to collapse. Peter Haff has described it as 'racing ahead like a forest fire'.[13] The vivid analogy gives pause for thought. In this situation, one thing quickly leads to another. Quite where all this is going to lead is anybody's guess. But one of those possibilities, now emerging, is the development of living machines, of flesh and blood.

Dawn of the xenobots

The technosphere has already blurred the dividing line between life and technology. We humans already walk the Earth augmented, many of us, by teeth filled with amalgamated mercury, silver, tin, and copper, hip joints of steel, titanium, plastic, and zirconia-toughened alumina, and electronic heart pacemakers. And while our basic biology remains more or less identical to that of our distant ancestors, the same cannot be said for broiler chickens, as we have seen: these newly modified organisms can now only exist, however briefly, within a technological support system. However, we can still more or less work out what is organism and what is machine.

That distinction, though, has just got more difficult.

The trouble with the machines we make is that, however complex, robust, decay-proof, and hardwearing they may be, each one is based on a single inflexible design that, by and large, cannot adapt and, when eventually worn out or broken, cannot self-repair, or reproduce itself. Living organisms, by contrast, may seem fragile by comparison with, say, a tank or fighter plane—but their ability to adapt, self-organize, repair their injuries, and produce new organisms has now been continually honed over more than 3 billion years. Could one therefore get the best (or worst, as some may think) of both worlds by making machines out of living matter, or—which amounts to the same thing—design organisms from scratch to fulfil specific purposes? A research team in the USA has just taken some steps to that very end,[14] and made what some people call xenobots.[15] These may be opening up a new kind of future somewhere in the space where technosphere and biosphere overlap that seems to be promising and disquieting in equal measure.

The team combined biological and computer expertise to create these new living machines. The 'xenobots' here has the ring of alien robots, but is derived from the raw material used, cells of *Xenopus laevis*, the clawed frog, which is one of the model organisms used by biologists. The particular cells used were embryonic stem cells—that is, those which still have the capacity to develop into any specialized kind of cell—but which were on the cusp of specializing into muscle cells capable of movement, and immobile skin cells. These were put into clusters about a millimetre across, and then sculpted into patterns: not random patterns, but ones which had been computer-designed to fulfil a particular purpose—to move in a certain fashion, for instance. With these blueprints as guides, the sculpting was then carried out by human hand, wielding tiny forceps to move cells into place, and a fine electrode to obliterate others. The researchers then sat back and watched their creations.

The xenobots functioned—but not always as expected. This is not like assembling a machine out of metal wires and microchips, where the properties are known, and the interactions can be precisely predicted. Here the machines are of living tissue, which has infinitely complex dynamics of its own. So, some of the xenobots 'walked' across the experimental surface, or manipulated objects, more or less as programmed. Others, though, developed new behaviours, attaching or moving around each other when they met, or forming new shapes such as holes within them, in effect improvising on or subverting the original designs.

Baby steps, perhaps, but the xenobots give pause for thought. The researchers point to possible uses; such bespoke organisms might be designed to deliver drugs within the human body, say, or clean up environmental pollution. How far might this new kind of science reach? In a final sentence, the possibility of producing such new forms of life with 'cognitive or computational functions' was mooted, to create a new kind of intelligence. Would it, one wonders, develop a mind of its own?

The technosphere, newly arrived on planet Earth, is clearly evolving at furious pace, as its ever more sophisticated machines interact with thousands of highly trained, imaginative, and innovation-focused humans at its leading edge. The possibilities, viewed this way, seem endless. Whether this new planetary experiment will survive long enough to produce a new pattern of life, though, is an open question. There is a hurdle to get through first, one that may be best viewed from a safe distance, before we examine it in ugly detail, close up.

The eternal technosphere

We humans are, biologically, just animals, creatures of flesh and bone. In his strikingly experimental novel William S. Burroughs called us *The Soft Machine*. In primitive form, we are not really designed for the kind of planetary immortality that palaeontologists call fossilization. After death, flesh rots quickly. Even bones, discarded on the landscape, begin to disintegrate within months, or are scavenged. This is why ancient hominid remains are so rare, despite being avidly sought as well-funded palaeontological teams scour the most likely fossil sites worldwide.

Burroughs' novel was influential in the most unlikely places. It introduced the phrase 'heavy metal', which was to become the banner for a whole genre of music, and for good measure, it was the direct inspiration for Soft Machine, a highly regarded jazz-rock band formed in the psychedelic 1960s. The band itself is durable (it is still going strong). Its personnel are, of course soft-framed in the standard biological way, but the ensemble cheats the grim reaper by recruiting new band members when needed.

The band's armoury, though, has a hardness and durability that will likely give it a rather greater chance of immortality, at least of the palaeontological kind. This Soft Machine has the standard rock band carapace of drums, guitars, saxophones, microphones, amplifiers, loudspeakers, swathes of electrical cables. Extending outwards, there is that realm, shared with other bands, of concert hall, of brick, concrete, steel, glass, and plastic. We can now watch electronic ghosts of the band, past and present, on our laptops, through the globe-encircling hardware that is several evolutionary levels beyond that with which we watched the primordial *Doctor Who* in our childhoods.

All of these artefacts are built to last. They are designed to resist microbes, termites, mice, spilled beer, over-enthusiastic fans, and careless roadies, to function faultlessly in sunshine and rain, drought, and frost. Steel is galvanized or chrome-covered. Wood is seasoned and varnished. Copper has one or two coats of that most indigestible of modern materials, plastic. Durability today is useful today. It is a very handy headstart, too, as regards the ultimate tomorrow of fossilization.

In nature, the fossils handed down to us from the deep past, from dinosaurs to trilobites to the most delicate of fossil leaves, are those which have escaped the highly efficient recycling processes of the biosphere. Dinosaurs, trilobites, and leaves in life are simply temporary stores of nutrients which, upon death, are passed on for equally temporary loan to the next transient generation of organisms. The rare scraps that spill over from this near-eternal cycle is what keeps palaeontologists in business.

The technosphere is much newer, and much less efficient at recycling. Today, it is growing and discarding its components as though there is no tomorrow. The Soft Machine's impressive hardware is durable—but it is not functionally everlasting. An amplifier blows, a guitar succumbs to over-enthusiastic use, a tape recorder becomes redundant—and the usual route from there is to join the growing mountains of discarded technology. These complex artefacts are designed to make profits for the manufacturer, not with recycling in mind. So, the Soft Machine's cast-offs will, more likely than not, make their own minor contribution to the truckloads that wend their way, each day, to our contemporary burial sites.

Burial is the first step towards fossilization, and modern landfill sites—particularly the most modern, highly engineered ones, are developed on a truly gargantuan scale. They are mostly invisible, placed out of sight and mostly out of mind, for the peace of mind of the people who live nearby—we will shortly visit one where the process is laid bare and gives a clearer and truer picture of what is going on.

There is a very good deal of overspill, too—the casual spills now all too visible in soils, rivers, beaches, and lakesides—while a far greater amount is accumulating invisibly in underwater sediments worldwide. Durability and rapid burial are the first steps towards fossilization, and to bear in mind the long future of today's artefacts, one may call them 'technofossils',[16] and some curious palaeontologists of the far future will find that one of the most visible parts of humanity's ultimate legacy to the planet will be strata loaded with the hyperdiverse petrified remnants of our trash:[17] flattened and carbonized plastic bottles, the impressions of aluminium cans—and among the rarities, perhaps the corroded and crushed remnants of an electric guitar (which would represent one of the more puzzling of those fossil enigmas-to-be).

This far future scenario is a safe, almost comforting one: many millions of years from now, these abundant new fossil assemblages will mostly be buried deep underground, far from where they can do harm (while some of the plastics will likely be stewing in the high subterranean heat and pressure to produce their own distinctive addition to future oil and gas resources). The examples exhumed by tectonic forces back to the surface will, petrified and rockbound, also provide little threat, while having—assuming there will be those curious far-future explorers to analyse them—a complex and abstract fascination.

Looked at from the here and now, though, the fascination is anything but abstract.

Used-up planet

For centuries the decrepit places where society has almost collapsed have lurked in our wilder imaginings, to be occasionally rendered in nightmarish visions. One of the masters of this art, the Flemish painter Hieronymus Bosch, lived in the Dutch city of S-Hertogenbosch. Like many medieval cities, S-Hertogenbosch was a cramped place of closely stacked buildings hiding behind its fortified walls. The fifteenth century was not a good time for S-Hertogenbosch. It suffered two great fires in 1419 and 1463. The young Hieronymus probably witnessed the second fire, and it may even have burnt down the house where he was born. It seems to have left a lasting impression on his psyche, and on his paintings, which often show depictions of hell that have a fire-scape of burning buildings in the background.

Around 1500, Bosch produced his most celebrated work of art, one that would become known as the 'Garden of Earthly Delights'. It is a large painting, a three-panelled triptych that is 2 metres high and nearly four metres long. Its left-hand panel shows a majestic Garden of Eden filled with many exotic animals including a giraffe, a lion, and an elephant, and just two people,

Adam and Eve. The central panel is a rambunctious rendering of a proliferating humanity enjoying all manner of 'Earthly delights' with a cavalcade of animals, some recognizable as goats and donkeys, whilst others are chimaeras. Everywhere these animals seem to be under the domination of people.

The images of the right-hand panel show the end game, in a hellish scene that warns what may happen when Earth's delights are consumed with impunity. Here humans fill the picture but are assailed by strange beasts on all sides, including a pig dressed as a nun, and cannibalistic and sadistic demons. Unconsciously, perhaps, Bosch retells human history, a world where once we walked amongst the beasts, that then became a world where we took mastery over the non-human. This grim final panel shows the kind of world towards which we may be heading here on Earth (and not in any hell of the afterworld) if we continue to consume unabated.

Such places already exist.

To the north of the Nairobi National Park, in the eastern suburbs of Kenya's capital city, is a vision of one of Bosch's hells, albeit brought right up to date. It is the Dandora rubbish dump. It lies just a few miles from the splendid giraffes and lions of the national park but is in reality a million miles away. Dandora is a poor suburb established in the late 1970s, ironically with money from the World Bank to provide a higher standard of housing. Something went awry in implementing this aim. Bracketed between the Nairobi River to the north and east, and the John Osogo Road to the south and east, the Dandora rubbish dump covers some 30 acres, about four times the surface area of the Acropolis in Athens. Each day about two-thirds of Nairobi's daily waste, some 2,000 tonnes, are delivered to the site on an endless conveyor of trucks from the city. This waste includes the surplus food from the airline flights that arrived at the international airport that morning. A row of mechanical diggers forms a kind of welcoming party as the waste arrives, as if in some scene from a post-apocalyptic *Terminator* movie. The food scraps are quickly picked over by hundreds of people who make a living from the dump. Some of the food is recycled for human use, but most ends up in bags for animal feed, sold for the equivalent of a few pennies. Metres-thick layers of rubbish have accumulated since the 1970s, including discarded syringes that poke out of the rubbish underfoot. The Nairobi River is slowly eating into this dump from the northern side and the debris then flows downstream into the city, carrying with it a cocktail of pollutants, quietly returning the rubbish to where it came from. When the rubbish is burned, the site reeks of noxious gases. Nevertheless, many people scratch a living here, assembling the rubbish into neat piles of plastic and metal that can be sold to make a living of a few dollars a month. If this scene was not Boschian enough, the waste is picked over by a small army of primeval-looking Marabou storks, each standing about a metre tall. Their bald heads identify them as carrion birds: being featherless, these heads are not clogged by blood and tissue as they ferret inside the carcasses of dead animals. To add to this menagerie, pigs roll in the mud and dirt, whilst gaunt men pick their way across the rubbish heaps, gnawing on discarded animal joints from downtown Nairobi restaurants.

Dandora is what planet Earth might look like when humans have *nearly* used everything up, when all of Earth's delights have been consumed, and nobody has thought to recycle anything for future generations. Dandora is not yet quite 'used up'. There is the plastic to recycle, and the tin cans, and discarded animal fat to be scraped off plastic bags and reused, and oxygen to breathe, though at night the acrid smoke from burning rubbish seeps into the homes of people living in the adjoining suburbs. Even so, this nearly used-up world of foul-smelling gases and discarded materials is still, by normal planetary standards, a marvellous gem in this universe. It is still a place of tough and resistant biology, infinitely more diverse ecologically than barren Mars, where wistful scientists dream of sending space missions to colonize a desert. Around the Dandora dumpsite, tobacco plants grow. These have strange chemistries preserved in their leaves that record the history of chemical contamination at the site. Metals like lead, cadmium, and chromium in the tobacco make smoking cigarettes made here a yet more hazardous business.[18] The bacteria that live at the Dandora dump are even stranger still. Many are pathogens, waiting to re-enter the nearby human population and then spread out into the city and beyond.[19] These include strains resistant to bactericides amongst the medical waste of the site. Like the humans that inhabit Dandora, these microbes show the ability of life to overcome the most seemingly hostile of environments. And one day some might mutate, escape, and cause widespread human tragedy, just as the COVID-19 virus has.

There are thousands of Dandoras around the world. In richer parts of the world, the disposal of rubbish usually takes place more secretively, protected from curious eyes by guards and fences, the rubbish—without the opportunity for the impromptu recycling that Dandora has—being buried in enormous plastic-lined pits. Out of sight and out of mind, perhaps. But many of these sites are just as much ecological time-bombs as Dandora, as changing conditions—the erosive power of a rising sea level, for instance—can exhume our garbage, and bring us face to face with it again.

Dandora is a vision of our planet used up. That world would be a kind of gigantic field of discarded artefacts, human feedstock, and depopulated seas, combined with a huge waste bin. It is not just an Easter Island with no trees, but a whole world devastated. It is a world where there is nowhere left for us to dissolve back into our ecosystem and live sustainably with it again. Instead, it is a world where humanity's impact rips through the Earth's resources like a forest fire out of control, via a technosphere running its course.

The problem of the technosphere's waste lies today, hence, and in the immediate decades, centuries, and millennia to come. It is now too indigestible, by far, for a diminished and ailing biosphere to cope with, and too abundant to be safely diluted by the hydrosphere and surface lithosphere. One might think of the plastic trash now drifting pretty much everywhere: much of it will sicken or kill the animals that mistake it for food, or that become trapped within it, before this new material is finally, safely, buried in some newly forming strata. Its waste gases are loading the atmosphere, too, to heat the planet and place further strain on its living organisms. And while the technosphere's own metabolism grows and ever more rapidly evolves, this evolution has not yet led to a capacity for recycling its own materials at anything like the scale needed to maintain its own future, or that of the future of the biosphere from which it so recently emerged, and which it now parasitizes.

If the parasite is not to kill the host, this is among the most pressing tasks for us component humans to attend to—while we still have some influence on this phenomenon, simultaneously marvellous and monstrous, that we have helped to emerge as a major planetary force. As the technosphere's grasp on the Earth System tightens, the consequences of its failure to recycle are growing.

But, there are ways of living that might help us to engineer human societies that live beneficially with nature. In the final chapter we ask how humans and the technosphere might coexist with what will remain of nature.

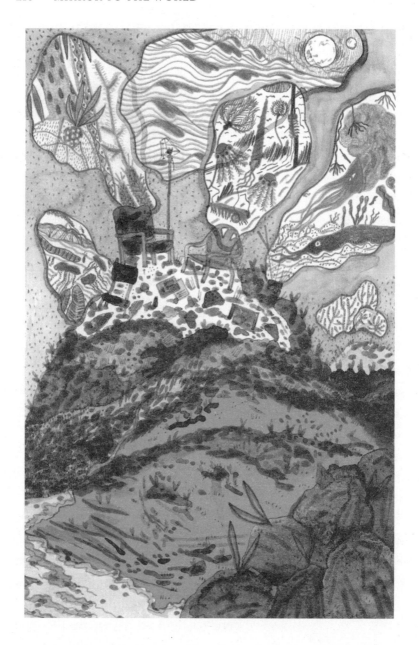

8

NO COUNTRY FOR
WILD APES

When the astronaut played by Charlton Heston rode across the beach towards the three-quarters buried Statue of Liberty in the 1968 movie *Planet of the Apes* there was a potential cinema audience of 3.5 billion people on Earth to watch him. Forty-three years later, when *Rise of the Planet of the Apes* was released, its potential worldwide cinema audience was 7 billion people, a doubling in half a human lifetime. And only a decade later there would be nearly 8 billion of us. In the final scene of the original movie, Heston's character barked out the apocalyptic words, 'You maniacs! You blew it up! Damn you! Goddamn you all to hell!' But, in that imaginary fortieth-century world, humanity was already nearly forgotten, and nature was restoring itself, with gorillas, chimpanzees, and orangutans now taking the leading roles.

Back in today's twenty-first century something different is afoot. Each day the number of human births exceeds the number of deaths by more than 200,000 and we add the equivalent population of Germany each year (over 80 million people) to the global population. It is difficult to think of all of those people, or even the number of people who live in one large city. Say the 13 million people of Tokyo, for example, whose numbers seamlessly transition into the major cities of Yokohama, Kawasaki, Chiba, and others to form a megalopolis of 38 million people, and whose urban margins can't really be fathomed even from an aeroplane flying into Haneda Airport. The much smaller metropolis of Birmingham in the West Midlands of England sits at the core of a metropolitan area of about 4 million people, or roughly the entire human population of Earth at 10,000 BCE. But even from the heart of this city it is still difficult to look out and fathom the 4 million West Midlanders amid the concrete sprawl. For most of us it is much easier to visualize smaller numbers. The old coal-mining town of Biddulph, in the North Staffordshire moorlands of England, nestles between two hills, Mow Cop and Biddulph Moor; its 20,000 inhabitants just about fill the valley floor, overlapping the surrounding hillsides in a few scattered houses and farms. Nevertheless, 10 Biddulphs will be added to the human population in the time it takes you to read this book. Or roughly double the total wild population of gorillas. When the numbers of other primates are compared to those of humans, the figures become even more telling.

There are a little more than 500 species of primate on Earth, mostly dispersed between tropical America, Africa, Madagascar, and South and South East Asia. Two-thirds of all these species occur in Brazil, the Democratic Republic of Congo, Indonesia, and Madagascar, making these the hotspots for biodiversity. More than half of all primate species are threatened with extinction as a result of agriculture, logging, the spread of ranching, and hunting.[1] Just a few major types of farming strongly drive this trend towards a primate mass extinction: those which supply our love of beef, our uses for palm oil and rubber, and our need for sugar, soybeans, and rice. And if these trends continue through the twenty-first century, agriculture will snatch away two-thirds of the land that primates need to survive, and this demise would put a stop to tens of millions of years of evolution. The problem is not just the loss of the primates themselves—let alone the countless other species that will disappear—but the ecological functions they perform as pollinators and distributers of seeds.

Our nearest living relative, the common chimpanzee, numbers just a few hundred thousand, or the population of one small human city like Buffalo in the USA or Nottingham in the UK. At the same time there are about 100,000 gorillas and just 70,000 orangutans. If more and more of the land they need to survive is grabbed by humans, then one view of Earth in 2100 is of a planet with no wild apes or other large wild mammals: one in which the remaining wilderness animals are restricted to zoos, and humans have become a dominantly urban species divorced from a degraded natural world turned (whatever remains productive) into feedlot. It is a world where the technosphere might emerge as a dominant system, an apocalyptic scenario with overtones of the artificially intelligent Skynet in the movie *Terminator*. Nothing epitomizes this impact on the planet more than the relationship we have developed with our closest relatives, the great apes. We are exterminating them by proxy, degrading the ecologies that support them.

The human ape did not always dominate the surface of the Earth, and even now amongst our nearest relatives, there are more species than immediately meet the eye, and thus a small chance to redeem the future. There are seven species of great ape in addition to us, that live in the forests of Africa and Asia: three species of orangutan—the Indonesian word means forest man—on the islands of Borneo and Sumatra; two species of gorilla in Africa, the eastern and western gorillas; and two species of chimp, the 'common' one and the gentle and promiscuous bonobos. Collectively, the total wild population of our seven cousins is about half a million. That makes all orangutans, gorillas, and chimps 0.006 per cent of living great apes, with the eighth great ape, humans, accounting for 99.994 per cent. And then there are the so-called lesser apes, or gibbons. Smaller in stature, but tail-less nonetheless, there are some 18 living species of gibbon dispersed through the forests of South and South East Asia. One of their number, 'Mueller's Bornean Gibbon', is counted as the second most abundant primate on Earth—after humans—with about 400,000 still in the wild. Most other gibbon species are severely endangered by the actions of humans, like the noisy and bellowing siamang, the largest of all gibbons, that lives in the forests of Indonesia, threatened by the seemingly unstoppable spread of oil palm.

No country for wild apes

You cannot easily avoid palm oil. It may be in your chocolates, biscuits, ice cream, cereals, fast food, soap, shampoo, toothpaste—and that's just for a start. Indeed, it may be in 50 per cent of all the items you buy from some supermarkets.[2] But like it or not, each time we buy these products we are edging orangutans out of the forests they occupy. Palm oil mostly comes from the African oil palm, and in the past two decades plantations of this tree have spread rapidly through the forests of Indonesia, Malaysia, and Thailand. Oil palm will only grow near the equator, especially in those places where the world's most biodiverse rainforest ecosystems grow. Within these forests many species are endangered by the spread of this plant.

Nowhere on Earth epitomizes this loss of natural forest to oil palm like Indonesia.[3] The rate of deforestation in this vast country of 13,000 islands and over 270 million people is dramatic, and the means by which this is achieved—largely by burning the landscape—has led to massive emissions of carbon dioxide to the atmosphere both from trees and from peatlands. Even in its more remote regions, like Borneo, deforestation is intense, and half of the lowland rainforest has been removed or degraded in recent years.[4] Here live some of the most biodiverse rainforests on Earth, rivalling those of the Amazon and Congo, with majestic trees rising 80 metres into the air. These forests contain perhaps more than 10,000 plant species. They are home to the vestiges of Borneo's surviving megafauna of elephants and rhinoceros. One animal in particular is emblematic of this forest loss.

Orangutans may once have lived widely in the forests of South East Asia during the Pleistocene,[5] their range extending as far north as southern China. For much of that time they likely co-existed in landscapes where small groups of early humans and other great apes like *Gigantopithecus* also lived. As the Earth's climate cooled into the peak of the last Ice Age about 25,000 years ago, the forests of South East Asia contracted, and with them the geographical range of the orangutans. They are found now only in small parts of the islands of Borneo and Sumatra. The three remaining species are encroached upon from all sides by humans.

In the northern part of the island of Sumatra, to the south of the giant Lake Toba, lie the forests of Batang Toru in the Tapanuli region. Orangutans have been observed in this region since 1939, but only recently, in 2017, were they recognized as a distinct species. These Tapanuli orangutans differ from their cousins, the Sumatran and Bornean orangutans, by their frizzier hair and smaller and flatter faces,[6] and by their distinctive and higher-pitched calls. They have never been seen at ground level and may live their entire lives in the treetops. Genetic evidence suggests that they separated from the Sumatran orangutans over 3 million years ago.

The Tapanuli orangutans live in the deep lush forest that clings precariously to the valleys and precipitous mountains of Batang Toru, in an area of about 1,000 square kilometres, about two-thirds of the space covered by Greater London. Fewer than 800 individuals have been counted, making them the rarest of all great apes, and critically endangered. These apes are already geographically divided into three separate groups, with only one having a viable population of about 500,[7] and their arboreal habit means they are threatened by infrastructure, like roads which they cannot cross. They cohabit with many other species that are endangered, including the Sumatran tiger, and all are assailed by the spread of agriculture, logging, mining, and the construction of a hydro-electric power station.

These are not the only threatened orangutans in Indonesia. Their cousins, the Sumatran orangutans, number about 14,000 in the wild. A more viable population than the Tapanuli orangutans perhaps, but most of the Sumatran orangutans occur in groups of a few hundred to a thousand and these are now limited to the north of the island, where the main threat to their habitats is the rapid spread of oil palm. Across the sea on the island of Borneo there may be more than 50,000 Bornean orangutans, but these too are threatened. Oil palm has spread like wildfire through this island, and now occupies 8.3 million hectares[8] (83,000 square kilometres), an area larger than Scotland.

Tapanuli orangutans may become extinct within a few decades of being recognized as a distinctive species. Their two more numerous orangutan cousins may follow into oblivion shortly afterwards if their habitats continue to be destroyed for oil palm plantations. History recapitulates, as humans do to orangutans what they did earlier to Neanderthals, appropriating the lands that these apes need to support themselves. This pattern is now emerging everywhere, and on the Asian mainland other apes are severely threatened.

Fighting back?

Orangutans at least have influential friends. One of the characters invented to populate Terry Pratchett's fantastical Discworld novels is the Librarian, guardian of the Unseen University's academic (and dangerously magical) books, regular visitor to (and occasional brawler at) the Mended Drum tavern down the road—and an orangutan. Pratchett recalled later that this creation was meant as a mild literary joke, which took all of 15 seconds of his time (on another day, he said, it might have been an aardvark guarding the books). But the Librarian was to become, along with Granny Weatherwax, the Patrician, and Death, one of the enduring characters among the denizens of the Discworld (one librarian fan thanked Pratchett for 'raising the status of the profession'), for whom Pratchett conjured up a myriad of subtly different meanings expressed by the utterance 'Ook'.[9]

The Librarian changed Pratchett's life too. It got him to think about the animals themselves, and their rapidly diminishing prospects. He found there was an Orangutan Foundation and began to give it some of the money that the Librarian had helped earn. Drawn into its work, he became a trustee, and it turned out to be a 22-year association with a cause that he felt passionately about.

How does such an enterprise start in the first place? There are many such, large and small, around the world, typically started by people with unusual levels of empathy, determination, and powers of persuasion. Much fine work gets done. The Orangutan Foundation, for instance, patrols about a million acres of rainforest to try to preserve the habitat, releases orphaned orangutans into protected reserves, devises the means of allowing people

from all around the world to 'adopt' an orphaned orangutan
before it can be released, and works to promote education and
sustainable livelihoods for local people. It's magnificent work and
has been praised by such as Sir David Attenborough.

Pratchett's clear-eyed view of the realities,[10] though, indicates
the scale of the task, and the inexorable odds at play. The profits to be made from logging, both legal and illegal, and palm oil,
are enormous, and such big money can assemble larger and more
sharp-edged forces than can the orangutans and their protectors.
Illegal encroachers on to the national parks, who see them as just
another form of timber supply, carry machetes, while the rangers
trying to protect them are simply armed with words and reason. At higher bureaucratic levels, issues of national sovereignty
sensibilities, and of (now enormous) foreign investment make
negotiations about habitat conservation, as Pratchett put it, 'like
tapdancing on quicksand'. And all the while, the forests shrink,
and so do orangutan numbers. And orangutans are not the only
apes to be threatened by the excesses of humans.

The diminutive ape cousins of orangutans, and of us, are the
gibbons. They were once widespread in China, and their numbers likely ranged into the hundreds of thousands or more. Their
cries could be heard by travellers along the great Yangtze River,
and their domain extended as far north as Gansu Province near
the border with Mongolia, and as far east as where the land meets
the East China Sea. An ancient Chinese text, the *Chu Ci*[11] records,
'The deep, gloomy forest is where gibbons and snub-nosed monkeys live.'[12] And for millennia the cries of these gibbons seeped
into the Chinese consciousness, finding their way into literature
and poetry.[13] Early writings noted gibbon songs, and the different colours of males and females, and that grown men could be
moved to tears by their mournful cries. Chinese culture admired
them as noble and graceful in the treetops, and as celestial, being

able to transmute from gibbon to human.[14] This blurring of the genetic space between humanity and other primates is common to many other cultural traditions.

Somehow, the economic progress of the late twentieth century blindsided this ancient reverence of gibbons, as though a wave of collective amnesia swept through people. This has—at least temporarily—benefitted the technosphere and its symbiotic humans, who have extended their control over the landscape, seeding oil palm and rubber plantations where pristine forest once grew. But the impact on the gibbons has been profound. Just a few vestiges of four species remain in China, none numbering more than the population of a small Chinese village.[15] One of the survivors is the Hainan black crested gibbon—the males have black fur, whilst the females are a luxuriant honey brown. In the 1960s there may have been as many as 2,000 individuals in the wild. 'As many' may seem an odd statement here, written about an ape whose numbers approach 8 billion individuals, but in the ensuing years the population of the black crested gibbon plummeted. In that time, much of Hainan's natural forest ecosystem has gone, felled by logging or cleared to make way for rubber plantations, and with it over 95 per cent of the gibbon's natural habitat, with all of the plants they need to sustain themselves. There are perhaps a few tens of individuals of this species still living in the wild. Their numbers represent no more than one extended human family.

Even so, the black crested gibbon has fared better than the northern white-cheeked gibbon. These gibbons survive in very small numbers over the border in northern Vietnam and Laos. They live in small family groups, sleeping in high branches where they cling to each other as they sleep. The last stand of these gibbons may be the Pù Mát National Park in Nghệ An Province of Vietnam. Here, on the high slopes of the mountainsides there are about 450 white-cheeked gibbons. They share these slopes with many other endangered animals like the Annamite striped rabbit, Indochinese tiger, and Asian unicorn, a relative of the antelope. As recently as the 1960s there may have been upwards of 1,000 northern white-cheeked gibbons living in China's Yunnan Province too, but none have been seen there for more than a decade.

As the technosphere ramifies across the landscape with its ever-increasing influence, are there any places were monkeys, apes, and humans might cohabit successfully?

The wolves amongst the monkeys

To the north of the capital city of Addis Ababa, on the Guassa Plateau of Ethiopia, monkeys and wolves have struck up a bargain, one that shows that mutually beneficial relationships can grow between very different animals. The plateau, formed from massive outpourings of basalt lavas several million years ago, rises to altitudes above 3,000 metres, where an Alpine-like grassland envelopes the western edge of the Great African Rift Valley. Here there is a rich diversity of wildlife including hyenas, leopards, wolves, and Gelada monkeys. The monkeys are not considered endangered, despite their population halving to about 200,000 individuals over the last half century. They are large monkeys, about a third of the average human weight, and males bear a distinctive hourglass-shaped red patch of skin on their chests that gives them the colloquial name of 'bleeding heart monkeys'. They live almost entirely by feeding on grass, an unusual diet for monkeys, supplementing this with flowers and seeds. Single males may live within herds of more than a thousand.

Amongst the geladas, Ethiopian wolves can be seen hunting for small rodents. This wolf—unlike the geladas, for now—is very much an endangered species, with perhaps only 400 in the wild, and like the geladas its range is limited to these highlands. The wolves and the geladas have formed a symbiotic relationship. The wolves do not hunt the geladas, unlike the wild dogs that patrol this area, and though the geladas keep an eye out for the presence of the wolves they appear not to fear them. The wolves appear to benefit from the relationship by doubling their success rate hunting small mammals. Though no one is quite sure how this happens, it may be that the very presence of the geladas flushes out small rodents from their subterranean passages.[16]

The wolves and monkeys signal ways in which very different animals can cohabit. Such relationships have also existed between humans and primates in many places and at many times, from the veneration of monkeys in sacred Mayan texts[17] to the Hindu god Hanuman who is worshipped in the form of a divine monkey. His temples and those of many other deities have teemed with monkeys for centuries. The monkeys forage for food supplied by the human visitors, and from the rubbish they discard. Some monkeys like langurs are even seen as descendants of Hanuman and are revered. But as pressures from population growth have encroached on temples, and the surrounding land for forage has diminished, monkeys have taken to raiding crops and households, and human–monkey conflicts have developed. The problem then becomes one of identifying spaces which can be shared for mutual benefit.

Around the world, over 50 species of primates are known to feed in plantations. Sometimes their presence is viewed with great suspicion by local farmers, and they are persecuted as crop stealers, hunted for bushmeat, or taken for the illicit pet trade. Sometimes these interactions mean that the primates catch human diseases or pass on diseases to us. But often the monkeys can fulfil their natural roles as seed dispersers, helping to stabilize forest ecosystems in the hinterland of plantations, to fertilize crops through their poo and, because they also forage for insects, preventing the defoliation of plants.[18]

Even in cities, monkeys and people can live together successfully. In the suburbs of the southern Brazilian city of Porto Allegre is the district of Lami where the urban and natural intertwine, with trees and powerlines sometimes dangerously close. Brown howler monkeys sometimes strayed onto the power lines and, attempting to support themselves by grasping across two cables, were electrocuted.[19] The local people sought to prevent this, and the powerlines were woven together and insulated, whilst at the same time monkey bridges were constructed. It was the locals who decided where the bridges would be built by monitoring where fatalities took place. Other animals like opossum and porcupines now use the canopy bridges, interconnecting forest fragments and helping to preserve ecosystems.[20] The bridges of Lami show that with some thought, humans and primates can cohabit.

The thick woods

It is not only the natural landscapes and forests occupied by apes, gibbons, and monkeys that have become imperilled from human encroachment. Since the 1980s many landscapes have faced mixed fortunes. Whilst forest cover has increased overall in the boreal, temperate, and sub-tropical regions of the Earth, many areas of the tropics have seen loss.[21] Much of this forest loss is for the short-term benefit of commerce[22]—oil palm in Indonesia, cattle ranching in Brazil. Overall, something of the order of half the trees that once grew on the land have been cut down by people over the past 10,000 years.[23] Logging and destruction of natural habitats now takes place at a global scale, so much so that many ecologies have been pushed to the brink of extinction, leading biologist E.O. Wilson to argue for a world that is half wild, half saved for nature,[24] if humans are to avoid causing a sixth mass extinction.

On the land, some 846 distinctive types of biological assemblage are recognized as 'ecoregions'—smaller geographical entities than biogeographical realms—each being defined by a distinctive flora and fauna that is controlled by the landscape, geology, and climate of its particular region, rather than by its geographical isolation. Of these ecoregions, 98 have at least half of their land set aside for nature, and for a further 541 this 50 per cent target is considered at least achievable, or there are sufficient areas of intact natural habitat that restoration is possible. But nearly a quarter of all ecoregions—207 of them—are identified as imperilled, with close to little or no natural habitat remaining, and where 96 per cent of the land is converted to human use.[25] These are often densely populated regions, with intense pressures from deforestation, intensive agriculture, urbanization, and pollution. But even in supposedly more remote regions, like the island of Borneo, forested ecoregions have become vulnerable, with potential catastrophic loss of biodiversity.

So, is it still possible to rewild half of the landscape? Should one even try? What does one do with the human communities that live on the areas to be rewilded, for instance? These kinds of questions have raised sharp debate about the morality of such an aim,[26] for they bring with them issues such as how equally humans are to blame for biodiversity loss, and who and what should be sacrificed along the road towards restoration.

If such difficult questions can be negotiated, rewilding would need a concerted effort to document areas where pristine, or near-pristine ecologies still exist, where the native species can be protected and nurtured so that viable populations of species can be maintained, and where natural evolution can continue, unimpeded by humans. But, perhaps most importantly, rewilded landscapes need to be large, to give enough space for animals like wolves, bison, and lions to migrate through, and to interconnect across climate zones, so that animals and plants can migrate as the climate changes. Such approaches are being developed, even at a continent-wide scale in some regions like the United States. They might provide some hope for the future if the problems that surround them can be solved.

Just over a third of the United States is covered by forest, but prior to the arrival of Europeans about half of the area was forested, separated by prairie lands, wetlands, and deserts. A quarter of a billion acres of woodland have been cut down in the USA—primarily for agriculture to feed a growing human population and their livestock—but about three-quarters of the forest cover remains, and many areas of this, like the forests of the western United States, preserve wilderness. Other regions are much more degraded. By the late nineteenth century as much as half of the eastern woodlands of the USA had been felled, with the subsequent loss or near-extinction of indigenous species like the green-feathered Carolina parakeet, one of only two parrots that ranged into the United States (the other species, the thick-billed parrot, survives in Mexico, though in small numbers). The Carolina parakeet was abundant in the forests of the east, living in large noisy flocks of a few hundred, and ranging as far north as New England. It went extinct in the wild possibly as early as 1910; its primary habitats degraded, and it may have succumbed to disease.

In the north-eastern corner of the USA, on the border with Canada, is Maine, the largest of the New England states and naturally biodiverse because it straddles the temperate to boreal climate zones. When European settlers arrived here during the seventeenth century they observed 'infinite thick woods' that were largely undisturbed by the indigenous Algonquian peoples. But, two centuries later, significant areas had been felled and most of the ancient woodlands have now gone, replaced by managed woodlands that are encroached upon by humans. These forests are changed: many of the ancient trees have gone, and with them much of the original biodiversity, including cougar, mink, wolves, and caribou. And because many of Maine's surviving mammal species are 'habitat specialists' with narrow ecological ranges, they are susceptible to local extinction. Non-native species have also progressively made their way into the local fisheries.

Here and in other regions of the eastern seaboard where much of the original ecosystem is degraded, the idea has emerged of constructing wildlife corridors[27] that connect the existing patches of wilderness and make highways for animal life to migrate through. The idea is not a new one. Three years after the last Carolina parakeet died in the Cincinnati Zoo in November 1918, a visionary forestry graduate from Harvard called Benton MacKaye was outlining ideas for an Appalachian Trail[28] that would connect together a 'belt of under-developed lands' through the Appalachian Mountains from Mount Mitchell in the south to Mount Washington in the north (in 1925 the trail was shown as running for over 2,000 miles, from Mount Katahdin in Maine, to Stone Mountain in Georgia).[29] MacKaye did not conceive this as a wilderness for the protection of wildlife, but instead as reconnecting people with nature to improve their well-being, and of developing possibilities of locally managed forestry and agriculture that could sustain rural communities along the trail.

The Appalachian Trail may have developed out of a philosophy that sought to reconnect people with nature, but the idea of long geographical corridors that connect natural habitats—irrespective of their political context—was radical. This was nature reasserting itself over human boundaries, albeit through the writing of a human. From this grew the concept of an 'eastern wildway' following much of the Appalachian Trail, and the wildlands network now gathers information on forests from the lands between the Gulf of St Lawrence to the Gulf of Mexico, traversing many ecoregions and climate zones, and uniting many organizations with a common aim of preserving and restoring the wildlife of this region. This began two decades ago with an examination of the forests of Maine. It takes the viewpoint of the animals and plants that live within the ecosystem to ask what kind of natural range they need to survive. The wildlands project is developing the idea of other wildways, that connect the Taiga forests of Canada, those of the Pacific seaboard, and the Rockies.

Growing garden cities

In landscapes where vast and wild forests remain, as in the USA, there is always something to build on. But how does one rewild a landscape where all of the pristine forest is degraded, and where there are many more people? Many such landscapes now exist, where almost no forest remains. Instead, there is an urban sprawl of concrete and tarmac welded together with the farmland that supports the city-dwellers. But there are clues in ancient place names, and in relict flower meadows that still grow, that formerly the land was filled by trees.

What might one do for now, about this kind of neatly arranged and increasingly globalized conformity? Well, the conformity part might be subverted a little. If left to run wild, our manicured gardens will not immediately return to their natural forest state, as the starting biological ingredients are now so different. Nevertheless, we can follow the spirit of that visionary we met earlier on, Robert Hart, and of his Shropshire garden. He hoped that we would use our gardens to rewild, to reverse what he saw as a 'suicidal trend' towards deforestation, which in the two decades since those words were written has eaten into the forests of South America, Africa, and South East Asia. In Hart's garden stood a rowan tree, and in the summer and autumn of 1994 he observed how a nasturtium and a runner bean entwined the tree, holding it as if in an embrace.

If Hart's philosophy was followed, what a different place a city could be, with its gardens linked together as a network through which native plants could creep and grow, and then grow into municipal parks and on into the countryside. The different shapes and sizes of gardens, and the surfaces of buildings associated with them, could be engineered to foster natural biodiversity, and not to shut out nature. Buildings could be intertwined with plants just like Robert Hart's rowan tree, and people could walk through forests of green, not grey, concrete blocks on their way to work. This would be a much fairer city too, not just for nature, but for those people who live with little or no access to gardens. This absence of green space became acute for many households during the lockdowns of the COVID-19 pandemic.[30] Robert Hart would have despaired at this. He saw gardens as our point of connection with nature, and that through them we could learn to live in a better relationship with other species.

So, imagine, for a moment, a scene from a different twenty-third century, where cities with gigantic skyscrapers are derelict places now visited as curiosities, just as we walk through the ruins of Palenque or Tikal today, and wonder how the mighty Mayan civilization met its demise. Instead of skyscrapers, the cities of the twenty-third century merge seamlessly into the landscape around them, so much so that they are difficult to see. The buildings are habitats for a multitude of species; they capture and recycle water, and even work to regulate the local climate. Life is thriving here, from the multitudes of birds roosting and feeding in the buildings, to the herds of goats happily feasting on the grassy knolls and vales that circumvent the cities' walkways. Buildings grow, they are covered with biological materials that generate energy, and they clean the atmosphere around them. And at the end of their lives they decay, fading away over many decades as the land gradually recolonizes their structures. In this way, a city might gradually flow across its landscape, just as a meandering river does, forming in some parts, and decaying in others.[31]

Regreening cities so fundamentally might sound like a pipedream, but the process may begin in a city near you quite soon. Singapore, for example, which once styled itself the Garden City, now likens itself to a city in a garden, bringing back great tropical trees into the heart of the urban metropolis. Along the elevated walkway of its large urban park, the Southern Ridges, it has planted several hundred native trees. Some, like the giant *Koompassia excelsa*, grow to over 80 metres tall. In the wild, this tree is often a home to Asian rock bees, which build large honeycombs in its branches. The bees gain shelter from the tree, but the tree gains from the bees, whose poo fertilizes the soil, ensuring its healthy growth. Robert Hart would approve of this mutualistic relationship.

Even at the small-scale level of the family garden or local park there are things that can be done to alleviate the pressure on insects and make space for biodiversity more generally. Leicester horticulturalist Jennifer Owen knew this. She was already a pioneer of garden biodiversity from the 1970s, when she returned to that city to take up a position at its university. Over the ensuing three decades she recorded over 2,000 species of insect visitors to her garden and documented many more plant and animal species that she wrote up in her book *Wildlife of a Garden: A*

Thirty-Year Study.[32] The organization Buglife today carries on this kind of work and makes it accessible to all.[33]

Such approaches to restoring wildlife will need to be scaled up if we are to prevent a sixth mass extinction. And this will require a wholesale rethink of our interaction with the land and oceans, and of the land of concrete and glass skyscrapers, of tarmacadam roads, and of disposable plastics. All must be completely refashioned if humans and the rest of nature are to survive in any kind of health. This task requires a deep rethinking of our attitude to how we consume the Earth's resources.

Remnant forests at the margins

Around the coast of the western approaches to Britain, many place names suggest a Viking presence. In the ending 'by' in the town of Tenby on the south-west coast of Wales, and further up the coast there is Ramsey Island. In old Norse, that name denoted the island of Hrafn. About midway between Ramsey and Tenby lies Skokholm, a small, flat-topped, grassy island, just a mile long and offshore from the Marloes Peninsula. The island is made almost entirely of rocks of the Old Red Sandstone, some 400 million years old, with some thin remnants of glacial clay left over from the retreating ice sheet that, very much later, once covered the Irish Sea. Today Skokholm is a bird sanctuary—for Manx shearwaters, oystercatchers, puffins, and many other species— lying just far enough from the mainland not to be visited often by people. The chattering birds give some sense of how the biosphere may have sounded to our ancestors, before the background drone of cars, factories, and cities drowned them out. But there, such comparisons would end, because the Norse name 'Skokholm' means wooded island, and the continued presence of bluebells and lesser celandine strongly suggests that trees once grew here. The island possessed no native land mammals, but rabbits were introduced by the Normans, and these still nibble at its vegetation. Skokholm is typical of many landscapes that have adapted to human influence over centuries. Its neighbouring island of Skomer is also cleared of woodland, and farther to the west, the remote outpost of Grassholm hosts 80,000 breeding gannets in the summer months, which cover its top with a white veneer of guano; formerly the island was green—hence 'Grassholm'—and used for sheep-grazing in the summer.

Skokholm is a small example of the wider deforestation of Britain, whose five ecoregions were all dominated by woodland. In its south-east, where rainfall is lower, lowland beech forest was abundant. Traditionally these forests supported a rich wildlife of hedgehogs, fox, red squirrel, otter, bats, and mice. Within its midst is stamped the heavily modified cityscape of London—itself covering over 1,500 square kilometres—and many other large towns have now concreted over its realm. Some fragments of this beech forest remain, such as the ancient woodland of Wychwood in Oxfordshire and in the New Forest of Hampshire and Wiltshire, but the Worldwide Fund for Nature (WWF) regards this ecoregion as critically endangered.

By far the most widespread of Britain's ecoregions was the Celtic broadleaf forest of oak, aspen, elm, birch, and ash. It rose to the north and west of the beech woodlands and extended through central and northern England and on into Wales and southern and eastern Scotland where the rainfall is heavier. Most of this ancient woodland has gone, its land taken by the great cities of central and northern England, by the tarmac roads and railway systems that interconnect them, and by the rolling agricultural fields that feed the city populations. Many of its indigenous animals, like the grey wolf and lynx, are long extinct.

In the far north of Britain grow the Caledonian conifer forests. These ancient woodlands are woven into the folk memory as an impenetrable barrier, as the place where the Roman Ninth Legion may have fallen in the early second century CE, ambushed by an ancient Celtic people or simply dissolved into its wilderness. Its pine trees are the hardy descendants of the first trees to appropriate the landscape of Britain after the retreat of the glacial ice. The Caledonian forests also contain oak, rowan, aspen, and silver birch. Despite their remoteness, they are much reduced, threatened by grazing and felling, and by replanting with non-native conifers. They cover only 2 per cent of their former area and yet provide habitats for birds that are not found elsewhere in the UK, like the flamboyantly plumed wood grouse, and the primordial-looking Scottish crossbill, which, as its name suggests, is found only locally.

Tiny fragments of two other ecoregions persist. The North Atlantic moist mixed forests populate the western and northern coasts of Scotland and Ireland with Scots pine, birch, chestnut, willow, and oak. Once these forests supported wolves, bears, and lynx. Foxes, badgers, and deer are still to be found, but this ancient ecology is now threadbare. And, clinging precariously to the sides of steep valleys in Wales and Scotland, is Celtic rainforest, in those parts of Britain where the rain seems to fall ceaselessly.

After such dramatic loss, can British woodlands be restored on such a densely populated island? In some small places people are attempting to revive them. On the Dundreggan Estate in the Highlands of Scotland[34] some 4,000 hectares of land are being returned to pristine Caledonian forest, and here wood grouse, red squirrel, golden eagle, and beaver are being reintroduced. This is not a simple task. The land has been grazed by sheep and deer for centuries, and much is bleak, degraded, and treeless. But using juvenile trees that are nurtured locally from the Highlands, gradually the forest is returning. Farther to the south, in the open hills of the Southern Uplands of Scotland, is the Carrifran valley. The landscape here is wild and beautiful, but the vegetation is mostly reduced to grassland by centuries of heavy grazing from sheep. The Carrifran valley was bare grass just two decades ago, with the occasional rowan tree, until the Borders Forest Trust intervened to buy the land, and began to restore the valley.[35] Now 600,000 native trees are growing, sourced from the local valleys.

In the wholly reconfigured landscape of the English Midlands, only small pockets of forest remain, here and there lovingly tendered by wildlife trusts. One of these is Cloud Wood in Leicestershire. It is just 33 hectares in extent. Woodland has been growing here since at least the time of the Doomsday Book in 1086, and formerly it may have covered as much as 240 hectares. Over the intervening centuries the forest cover changed dramatically, as areas were felled for farmland, and later the wood was used for burning locally quarried limestone to make lime. The woodland was clear felled in the 1940s, and to add ignominy the site was used for dumping quarry waste until the 1990s. Despite all this, the landscape retained its capacity to regenerate, and a diverse woodland has resprung, with over 200 species of plants. Evidence of the old and ancient woodland is found in its diverse flora, which includes the beautiful yellow pimpernel, the wonderfully named hairy wood-rush—a grass with tiny flowers—and the pendulous sedge, with its long flower spikes that sway gently in the breeze.[36] Cloud Wood shows the amazing capacity of the natural world to regenerate, if only it is given a chance.

These reforestations and regenerations are, at present, just the beginning, and much more of the landscape could be returned to a natural state. The forests of the English Midlands are small compared with the much larger area of grouse moor in England,

Wales, and Scotland, deer-stalking terrain in Scotland, or cereal crops in the UK. How much of the UK landmass could be converted back to natural forest? Fully 73 per cent is farmed at present, but that given over to pasture for feeding cattle and sheep yields a smaller overall percentage of food, because of the inevitable loss of energy through the food chain, while the grouse- and deer-country might be regarded as recreational for a small section of society, rather than a serious food resource. If our desire for eating meat could be reduced—or if it could be sated by artificial meat[37]—then trees and forests could once again flourish, carbon would be removed from the atmosphere and locked into the skeletons of trees, and much more space would be available to wildlife. Farmers would need to be rewarded for making a better future for all of us. But the trees would once more grow around us, and the small forests that are being restored now could eventually be joined to make wider pathways for wildlife to flourish across the landscape.

Interventions beyond the land

Where land meets the sea, much can be made of the sustainable relationships that indigenous peoples still build with their environment.[38] Their experience of fish is not fashioned by wholesale markets, but by a need to sustain what is available in their environment, and only to take what is needed. The Ts'myseen peoples of the northern coast of British Columbia, for example, have fishing practices that have survived for thousands of years.[39] They fish by pole and hook, allowing the hunter to select only the bigger fish. And they build semi-circular traps of boulders and stones that capture the autumnal migration of salmon as they return to spawn in their birth rivers. And where they use nets, these are deployed only to capture what is needed. They take only enough to provide the community with food and items for trade, and in this way secure their livelihoods and the life around them.

How can we extend this kinship with nature to enhance the movements of fish and crustaceans, the growth of mangroves, seagrass, and coral, and connect this with the land-based ecosystems that they also support? We can start where humans have traditionally cohabited with sea creatures, at the boundary between the land and the sea.

At the edge of many tropical seas grow mangrove forests. The gnarled and twisted submarine roots of these salt-tolerant shrubs and trees provide nurseries for fish, and habitats for arthropods, molluscs, and sea urchins. Above the waterline, insects, birds, and mammals thrive. Beyond their critical importance to coastal marine ecosystems, mangroves are also valuable to humans, protecting the land from storms and from sea level rise, and providing a habitat for seafood if carefully managed. But, half of the world's mangroves have disappeared, a result of pollution, clearing for agriculture and aquaculture, overharvesting for wood, and of damming and irrigation that changed the saltiness of coastal waters. Much of this loss has occurred over the past 50 years, with some 320,000 square kilometres of mangrove disappearing,[40] an area roughly equivalent to Italy.

One such place of loss is the island archipelago nation of Indonesia, home to the greatest extent of mangrove forests on Earth, though on the island of Java much of this has already been badly degraded. On the north coast of Java lies the ancient sultanate of Demak to the east of the city of Semarang. The sultanate was founded in the fifteenth century and was once the mightiest along the north coast. Now its coastline is vulnerable to a rising Java Sea. With excessive groundwater extraction causing land to subside, and shrimp fisheries spreading along the coast, many tens of thousands of people's livelihoods are threatened as the sea encroaches, a pattern that is emerging for coastal communities around the world. At the local scale this change is dramatic. Villages and houses that used to be several kilometres from the sea are now much closer or, as in the case of the small hamlet of Rejosari Senik, they have already been inundated.[41]

Future projections of sea-level rise suggest that the Demak coast could be inundated by flooding, up to 6 kilometres inland, by the end of the twenty-first century. As a response, the 'Building with Nature in Indonesia' programme has intervened, to try to prevent an impending disaster[42] and develop a sustainable coastline that could regenerate the mangrove and provide protection for the people that live there. Because the coastal ecosystem was already badly damaged it was impossible to simply replant mangroves. Instead, the coast is being engineered to trap sediment so that plants can re-establish naturally. Once this begins to happen aquaculture ponds can regenerate as the coastline stabilizes, providing a livelihood through crab and shrimp farming for Demak's seashore communities. The Demak project is a model that can be applied elsewhere along the coast of north Java, where up to 30 million people are threatened by rising sea level.

Mangrove ecosystems can coexist with aquaculture, where seaweed, molluscs and fish can be grown in cages within the mangrove waterways, without deforesting them. Other traditional, and sustainable, methods of farming in mangroves include the Gei wai ponds of China.[43] These are built among mangroves on mud banks and control the flow of water into them by sluice gates. This allows water, and shrimps, to enter at high tide, and instead of destroying the mangrove it is the very productivity of the ecosystem that feeds the shrimp and other life.[44]

Within shallow coastal waters, seagrass meadows also grow although, as with the mangroves, large areas have been lost.

Seagrass is not seaweed, but the descendant of land plants that returned to the oceans tens of millions of years ago. Seagrass has roots and flowers and seeds, and as with meadows on land provides food and shelter for many animals, from turtles to crabs and fish. It grows around the coasts of all continents, bar Antarctica, with the greatest diversity of species in the underwater meadows of tropical South East Asia. But the area of shallow seas it covers has been shrinking rapidly. Poisoned by pollution, dredged, or trawled, it is estimated that an area of seagrass covering the size of two American football pitches is lost each hour,[45] resulting in a global decline of 30 per cent since measurements began in 1879.[46] This is serious, not just because seagrass is a boon for biodiversity and supports inshore fisheries, but because it also stores carbon within its tissues and produces oxygen from photosynthesis that animals and microbes can use.

Around the coast of the UK the demise of seagrass has been very severe, with over 90 per cent of its meadows lost. This makes the few remaining places where seagrass still grows, like the Llŷn Peninsula in north Wales, vital resources for rewilding. It is providing seeds to the seagrass ocean rescue programme that is replanting a meadow off the coast of the Dale Peninsula in south Wales.[47] Here, at a carefully chosen site with the input of local people, so as not to disrupt fishing or moorings, a million seeds of the common eelgrass (*Zostera marina*) have been sown to replenish an area of two hectares.

The greening of the undersea meadows at Dale is a small step, one of many restoration projects of seagrass across the globe. On the western side of the Atlantic are the seagrass meadows of Chesapeake Bay. The Bay is 200 miles long and bordered by six American states. In its southerly part it is salty, but it becomes brackish and fresher towards the north, from the input of many rivers. Because of these salinity changes, many different kinds of seagrass grow, including eelgrass in the saltier waters of the south—the same species as at Dale. The major rivers that enter the bay also deliver industrial waste, with storm waters during extreme weather events washing in more pollutants. Over the years these pressures have led to the seagrass cover see-sawing. Nevertheless, restoration has helped to nurture 100,000 acres of seagrass, with ambitious plans for more.[48]

Beyond the restoration of coastal waters much more can be done to prevent the destruction of ocean life through mineral extraction, overfishing, and coastal development. Some of these pressures are so great that even our rapid intervention now may not, for instance, prevent the loss of much of the Great Barrier Reef as global temperatures rise. But, there are serious attempts to conserve biology in the high seas, through the establishment of protected ocean sanctuaries. One of the largest of these is Papahānaumokuākea (the name is a combination of the mother goddess Papahānaumoku and her consort the sky father Wākea), a marine sanctuary that lies to the west of the main Hawaiian Islands.[49] It is built around subsiding and ancient volcanic seamounts that extend for over 2,000 km from Nihoa in the east to Kure Atoll in the west, and it covers an area of the Pacific of over 1.5 million square kilometres. Here there is land, coral reefs, submarine seamounts, and open ocean in an array of interconnected ecosystems, where large populations of seabirds signal generally healthy seas. But even here, in its remotest parts, the sanctuary is not completely insulated from the technosphere. On the beaches of Kure atoll tonnes of plastic debris and fishing nets wash ashore, carried on the Pacific current, that can entangle sea mammals, turtles, birds, fish, and even lobsters.

The plastic pollution of Kure atoll reminds us that restoring ocean ecologies—and those of the land—will not work unless we, especially those of us in wealthy countries, change our patterns of consumption. It asks us to think each time we reach for a bag of frozen prawns in the supermarket. Where did those prawns come from, how were they farmed, and what environmental damage was done to the local environment? And how much energy was used to transport them to a supermarket aisle thousands of kilometres remote from their source?

The problems of the world's oceans may seem to overwhelm the immediate capacity of each of us to act, but there are some remarkable initiatives that—enabled through the easy access of the internet—allow all of us to make a difference. These include 'Seabirdwatch' with its over 10,000 citizen contributors, that monitors bird populations along spectacular coastal settings in the North Atlantic.[50] Healthy seabird populations signal a functioning marine environment, as the birds survive where seas are well stocked with fish. By using remote cameras to monitor the local bird populations throughout the year an army of citizen scientists pick through the images to identify important patterns of breeding. Equally remarkable is the project 'Fishing in the past'[51] that uses unconventional sources of information, like art, to piece together the abundance and diversity of fish from a time before systematic data were collected. Depictions of fish have a long history[52] and often can be identified to species level, as in the tombs of Egyptian pharaohs, where Nile perch, striped mullet, and tilapia are present. Such sources, and the very realistic depictions of fish in Dutch paintings of the Renaissance, provide information about the abundance and diversity of freshwater and marine fish through important intervals of climate change, such as the Little Ice Age. Gathering this information is a gargantuan task that needs the eyes of an army of volunteers to observe and identify. But it is one that can help quantify humanity's long impact on the oceans.

Rewilding the life within us

Out of sight, and beyond the landscapes and seas that we have imagined through this book, there is another habitat for life. One that is surprisingly diverse, and that protects and nurtures us. This is the life that lies within us, our own mini biosphere of trillions of cells that we shelter. Our bodies are colonies for these bacterial cells, and within the human gut alone there may be as many as 1,000 different species[53] together with fungi, archaea, and viruses. Our bacterial cohabiters may equal the number of our own cells (earlier estimates suggesting that our live-in bacteria might outnumber our own cells tenfold seem unlikely), but because of their tiny size, they only weigh about 200 grammes.[54] Most of this bacterial mass is in your gut.

Our microbiota has been co-evolving with us for millions of years and it likely developed from the foraging diet of mainly plant material—fruit, leaves, nuts, and tubers—of our distant ancestors. It protects us from attack from harmful pathogens, is beneficial to our digestive systems, and even seems to help our mental well-being. But sometimes components of our microbiota can harm us, being implicated in diseases such as bowel cancer. There is growing evidence that our modern way of living with calorie-rich, highly engineered foods, coupled with the effects of widespread antibiotics and the ultra-clean environments in which many of us live, is compromising the biodiversity of this ecosystem within us, and this loss might be implicated in the rise of chronic conditions like diabetes and asthma.

What was our ancient gut-beneficial diet like? Perhaps the best glimpse we have is that of the hunter-gatherer Hadza people in northern Tanzania. They occupy a landscape where ancient humans once trod, just a few tens of kilometres from both Olduvai

Gorge and Laetoli. The Hadza may have been in this landscape for tens of thousands of years, and genetically and linguistically are discrete from peoples who moved here much later. Only a few hundred of the Hadza continue to practise a nomadic lifestyle. Their diet consists of fruit, tubers, honey, and meat, collected with traditional tools like digging sticks and bows and arrows.

The gut biotas of those few human populations that still follow a foraging existence are considerably more diverse than those of humans eating a Western meat-rich diet; those of the Hadza even reflect the seasons of the year, as their food sources change.[55] The diets of many city-dwelling humans have caused some microbiologists to talk of an extinction of many types of bacteria, and of the need to rewild our guts, through a fundamental change in our diet back towards plant fibres. Each of us, therefore, has the capacity to help preserve biodiversity even at the intimate level of our own bodies. Returning to a more plant-rich diet, it seems, can protect the life within us, and us too.[56] And in reducing the burden on the land to feed cattle or on the sea to produce fish, it can also protect the Earth's biosphere.

Postscript: Leviathan and Gaia

As the Spanish Armada approached the English coastline in 1588, Thomas Hobbes was born—prematurely, as a result of his mother's fear of impending invasion. And fear, he later said, was the twin brother he was born with. It was a wry observation, and apt, for Hobbes later in life held a mirror to the human condition, in which reflection a monster appeared: the *Leviathan*. It was a key book in the development of political philosophy, but during his own life it brought rebuke, censorship—and, indeed, fear, into his life.

The Leviathan is powerful, cruel, and arbitrary—but it is also what prevents human life from being a constant war of everyone against everyone, or, in Hobbes' telling phrase, 'nasty, brutish and short'. Humans, he says, make a social contract for a sovereign to have absolute power over them, in exchange for security and stability. This blunt, transactional view—with no sugar-coating of sacred duty or God-given rights—was famously illustrated by a picture of a monarch wielding sword and crozier, his body made of hundreds of tiny human bodies. This, he said, is how a society needs be built, so as not to descend into chaos.

We are all today part of some Leviathan, though this monster, examined more closely, loses the sharp, clear outlines of Hobbes' vision. Indeed, there are different Leviathans: nation-states, multinational corporations, religious and political movements. And, rather like an electron is both a wave and a particle, their form cannot be pinned down. They are variously—depending on perspective and circumstance—physical bodies, sets of laws, patterns of exchange of matter and energy, means of communication, that are forever shape-shifting and forever impossible to fully grasp.[57] They have also grown hugely since Hobbes' time, bulked by steel and concrete, driven by fossil fuels, wrapped in plastic, communicating by computer and satellite, and fighting with high explosive and drones rather than sword and gunpowder. The single global Leviathan that emerges from these warring parts might well be, in at least one of its forms, the technosphere that we have explored in this book.

From the time of Hobbes onwards, the Leviathan has grown to stride across a landscape that has been assumed to be always there to nurture and support it. Everything was there, inexhaustibly, that was needed for a rich and full human life: food, drink, shelter, energy, materials, a stable climate. For a long time, that has been effectively so, and successions of human societies could focus simply on regulating their own lives through elaborate systems of laws, belief systems, economic plans, and the rest of the paraphernalia that fill—and direct—our lives.

That landscape, of forests and grasslands, bordered by bountiful oceans, had a life—and regulatory systems—of its own, well before the Leviathan came upon the scene. But to begin to grasp the totality of that life, though, was to take another three centuries after Hobbes. Life, that infinitely more ancient figure, was called the biosphere by Vernadsky in the early twentieth century, and Gaia by James Lovelock in the 1960s. We have followed in these pages just what a complex and ever-changing figure this is. It has always managed to right itself, for more than 3 billion years, true, to keep life going on this singular planet of ours—but it is not inexhaustible, nor invulnerable, and has suffered grievous injury in the past. The healing process afterwards can take millions of years.

Leviathan—which is, of course, a child of Gaia—is its latest challenge. It is a child grown enormous, and destructive. Ancient Gaia, under its sudden assault, has shrunk to half its size, while newcomer Leviathan has grown, now, in just the last few decades, to outweigh its ailing parent—and is still growing, like nothing else has ever done on this planet.

Will these two giants of our planet, one ancient and one utterly new, be able to live with each other, after this stormy and destructive beginning to their relationship? James Lovelock suggested that a wounded Gaia could exact terrible revenge on a delinquent humanity, and perhaps shake it off the Earth. That, alas, is becoming all too real a possibility. On a planet undergoing a high fever, microbial colonies and jellyfish may well prove the more resilient, for all of the Leviathan's current might.

Maybe that stand-off can yet be avoided, as we try to care for our cosmic oasis. We are all, of course, infinitesimal parts of both Leviathan and Gaia. And, although infinitesimal, we have come to understand both of them better than ever before, and are not without some capacity to nudge them towards safer ground. Maybe we can help shape their relationship away from harm, loss, fear, anger, and retribution, towards understanding, tolerance, support—and perhaps even love.

END NOTES

CHAPTER 1

1. The possibility of microbial life high in Venus's atmosphere, where temperatures and pressures are similar to the surface of the Earth, is being investigated: M. Wall. 2020. 'Life on Venus? Breakthrough initiatives funds study of possible biosignature', *Scientific American*, 16 September. https://www.scientificamerican.com/article/life-on-venus-breakthrough-initiatives-funds-study-of-possible-biosignature/.
2. Habitable exoplanets catalog. http://phl.upr.edu/projects/habitable-exoplanets-catalog (accessed November 2021).
3. Laetoli footprints. https://www.pbs.org/wgbh/evolution/library/07/1/l_071_03.html (accessed November 2021).
4. F. Détroit et al. 2019. 'A new species of *Homo* from the Late Pleistocene of the Philippines', *Nature*, **568**, 181–186.
5. Though for a deeply sympathetic perspective, see William Golding's 1955 novel *The Inheritors*.
6. J. Zilhão et al. 2017. 'Precise dating of the Middle-to-Upper Paleolithic transition in Murcia (Spain) supports late Neandertal persistence in Iberia', *Heliyon*, **3**(11), E00435. doi:https://doi.org/10.1016/j.heliyon.2017.e00435/.
7. J.F. Hoffecker. 2018. 'The complexity of Neanderthal technology', *Proceedings of the National Academy of Sciences*, **115**, 1859–1961. https://www.pnas.org/content/115/9/1959/.
8. B.L. Hardy et al. 2020. 'Direct evidence of Neanderthal fibre technology and its cognitive and behavioural implications', *Scientific Reports*, 10, article 4889. https://doi.org/10.1038/s41598-020-61839-w/.
9. K. Hardy et al. 2012. 'Neanderthal medics? Evidence for food, cooking, and medicinal plants entrapped in dental calculus', *Naturwissenschaften*, **99**, 617–626. https://link.springer.com/article/10.1007/s00114-012-0942-0/.
10. C. Shipton et al. 2018. '78,000-year-old record of Middle and Later Stone Age innovation in an East African tropical forest', *Nature Communications*, **9**, article number 1832.
11. Described in *A Journey through the Ice Age* by Paul Bahn and Jean Vertut, 1997.
12. This work is published by D.L. Hoffman et al. 2018. 'U-Th dating of carbonate crusts reveals Neanderthal origin of Iberian cave art', *Science*, **359**,

912–915. The age of the cave art, calculated at 64.8 ka, has been questioned by L. Slimak et al. 2018. 'Comment on "U-Th dating of carbonate crusts reveals Neanderthal origin of Iberian cave art"', *Science*, **361**, eaau1371, who propose an age of 47 ka.

13. J.A. Serpell. 2003. 'Anthropomorphism and Anthropomorphic selection—Beyond the "cute response"', *Society and Animals*, **11**, 83–100.

14. J. Peters et al. 2014. 'Göbekli Tepe: Agriculture and domestication'. In C. Smith, ed., *Encyclopaedia of Global Archaeology* (New York: Springer). doi: https://doi.org/10.1007/978-1-4419-0465-2_2226/.

15. M. Gilbert et al. 2018. 'Data Descriptor: Global distribution data for cattle, buffaloes, horses, sheep, goats, pigs, chickens and ducks in 2010', *Scientific Data*, **5**, 180227. doi: 10.1038/sdata.2018.227.

16. J. Coley. 2000. 'Roman Games: Playing with Animals'. In *Heilbrunn Timeline of Art History* (New York: The Metropolitan Museum of Art). http://www.metmuseum.org/toah/hd/play/hd_play.htm (accessed November 2021).

17. D.L. Bomgardner. 1992. 'The trade in wild beasts for Roman spectacles: A green perspective', *Anthropozoologica*, **16**, 161–166. D.L. Bomgardner. 2000. *The Story of the Roman Amphitheatre* (London: Routledge).

18. The story is vividly told in Stephen Greenblatt's book *The Swerve*.

19. A. von Humboldt, and A. Bonpland. *Personal Narrative of Travels to the Equinoctial Regions of America, during the Year 1799–1804*. Vol. I, Chapter 6. Translated and edited by Thomasina Ross. https://catalog.hathitrust.org/Record/004977751/.

20. A. Ghilarov. 1998. 'Lamarck and the prehistory of ecology', *International Microbiology*, **1**, 161–164.

21. His extraordinary life is well captured in a magnificent biography: K.E. Bailes. 1990. *Science and Russian Culture in an Age of Revolutions: V.I. Vernadsky and his Scientific School* (Bloomington and Indianapolis: Indiana University Press).

22. V.I. Vernadsky. 1998. *The Biosphere* (complete annotated edition: forward by Lynn Margulis and colleagues and introduction by Jacques Grinevald) (New York: Copernicus (Springer-Verlag)).

23. W. Steffen et al. 2020. 'The emergence and evolution of Earth System science', *Nature Reviews Earth & Environment*, **1**, 54–63.

24. Y.M. Bar-On et al. 2018. 'The biomass distribution on Earth', *Proceedings of the National Academy of Sciences*, **115**, 6506–6511.

25. T.M. Lenton et al. 2020. 'Life on Earth is hard to spot', *The Anthropocene Review*, **3**, 248–272. https://doi.org/10.1177/2053019620918939.

CHAPTER 2

1. Darwin married his cousin Emma Wedgwood.
2. C. Menta. 2012. 'Soil Fauna Diversity—Function, Soil Degradation, Biological Indices, Soil Restoration'. In G.S.A. Lameed, ed., *Biodiversity Conservation and Utilization in a Diverse World*. Intech Open. doi: 10.5772/51091.
3. T.W. Crowther et al. 2019. 'The global soil community and its influence on biogeochemistry', *Science*, **365**, 772.
4. M.I.J. Briones. 2018. 'The serendipitous value of soil fauna in ecosystem functioning: The unexplained explained', *Frontiers in Environmental Science*, **6**, 1–11.
5. N.E. Stork et al. 2015. 'New approaches narrow global species estimates for beetles, insects, and terrestrial arthropods', *Proceedings of the National Academy of Sciences*, **112**, 7519–7523. https://doi.org/10.1073/pnas.1502408112.
6. I. de S. Oliveira et al. 2012. 'A world checklist of Onychophora (velvet worms), with notes on nomenclature and status of names', *ZooKeys*, **211**, 1–70.
7. Though we note that earlier scholars had also considered such questions and made notable advances in this field. See for example, A.H. Malik et al. 2018. 'An untold story in biology: The historical continuity of evolutionary ideas of Muslim scholars from the 8th century to Darwin's time', *Journal of Biological Education*, **52**, 3–17.
8. A. Pressman et al. 2015. 'The RNA world as a model system to study the origin of life', *Current Biology*, **25**, R953–R963.
9. We discuss the distinction at length in our book *Skeletons: The Frame of Life* (Oxford University Press, 2018).
10. The story is told in N.R. Pace et al. 2012. 'Phylogeny and beyond: Scientific, historical and conceptual significance of the first tree of life', *Proceedings of the National Academy of Sciences*, **109**, 1011–1018. Woese's paper is C.R. Woese, and G.E. Fox. 1977. 'Phylogenetic structure of the prokaryotic domain: The primary kingdoms', *Proceedings of the National Academy of Sciences*, **74**, 5088–5090.
11. L.A. Hug et al. 2017. 'A new view of the tree of life', *Nature Microbiology*, **1**, article number 16048.
12. C.R. Woese. 2002. 'On the evolution of cells', *Proceedings of the National Academy of Sciences*, **99**, 8742–8747.

13. L. Sagan. 1967. 'On the origin of mitosing cells', *Journal of Theoretical Biology*, **14**, 225–274. For a later overview, see J.M. Archibald. 2015. 'Endosymbiosis and eukaryotic cell evolution', *Current Biology*, **25**, R911–R921.

14. e.g. Z.R. Adam et al. 2017. 'A Laurentian record of the earliest fossil eukaryotes', *Geology*, **45**, 387–390.

15. L. Yin et al. 2020. 'Microfossils from the Paleoproterozoic Hutuo Group, North China: Early evidence for eukaryotic metabolism', *Precambrian Research*, **342**, 105650.

16. N.J. Butterfield. 2000. '*Bangiomorpha pubescens* n. gen, n. sp.: Implications for the evolution of sex, multicellularity, and the Mesoproterozoic/Neoproterozoic radiation of eukaryotes', *Paleobiology*, **26**, 386–404.

17. N.J. Butterfield. 2011. 'Animals and the invention of the Phanerozoic Earth System', *Trends in Ecology and Evolution*, **26**, 81–87.

18. P.F. Hoffman et al. 2017. 'Snowball Earth climate dynamics and Cryogenian geology-geobiology', *Science Advances*, **3**, e1600983.

19. N.J. Butterfield. 2015. 'The Neoproterozoic', *Current Biology*, **25**, R859–R863.

20. We tell the story of the nature and history of this biological hardware in *Skeletons: The Frame of Life* (Oxford University Press, 2018).

21. D.E.G. Briggs. 2015. 'The Cambrian explosion', *Current Biology*, **25**, R865–R868.

22. R. Sender et al. 2016. 'Revised estimates for the number of human and bacteria cells in the body', *PLOS Biology*, **14**(8), e1002533.

23. L. Ramsköld, and X. Hou. 1991. 'New early Cambrian animal and onychophoran affinities of enigmatic metazoans', *Nature*, **351**, 225–228.

24. S. Conway Morris. 1999. *Crucible of Creation* (Oxford, New York: Oxford University Press).

25. The story is nicely told by Jack Sepkoski's son, David, in: D. Sepkoski. 2005. 'Stephen Jay Gould, Jack Sepkoski, and the "Quantitative Revolution" in American Paleobiology', *Journal of the History of Biology*, **38**, 209–327.

26. M.J. Benton. 2001. 'Biodiversity on land and in the sea', *Geological Journal*, **36**, 211–230.

27. He also replotted Sepkoski's diagram of the history of oceans for all organisms, not just animals: it did not substantially change its shape.

CHAPTER 3

1. H.W. Bates. 1843. 'Notes on Coleopterous insects frequenting damp places', *The Zoologist*, **1**, 114–115.

2. J. Planck. 2013. 'William H. Edwards and the Amazon', *Greene County History*, **37**, 11–16. http://docplayer.net/52232472-William-h-edwards-the-amazon. html/.

3. G. Beccaloni, and C. Smith. 2015. 'Biography of Wallace'. https:// wallacefund.myspecies.info/content/biography-wallace (accessed October 2021).

4. J. Van Wyhe. 2012. 'Alfred Russel Wallace. A biogeographical sketch'. http://wallace-online.org/Wallace-Bio-Sketch_John_van_Wyhe.html (accessed October 2021).

5. J.T. Costa, and G. Beccaloni. 2014. 'Deepening the darkness? Alfred Russel Wallace in the Malay Archipelago', *Current Biology*, **24**, PR1070–R1072.

6. P. Whitefield. 1996. *How to Make a Forest Garden* (Permanent Publications, Hyden House Limited).

7. Invasive non-native plants. https://www.rhs.org.uk/advice/profile? pid=530 (accessed November 2021).

8. Allan Mills' remarkable detective work on this phenomenon is described in J. Zalasiewicz. 2007. 'The spirit of biodiversity', *Palaeontological Association Newsletter*, **64**, 20–27.

9. https://mortaltree.blog/2014/02/02/robert-harts-forest-garden/ (accessed November 2021).

10. This is the process by which the Earth's surface lithosphere is divided into a series of discrete plates, which jostle for position at the surface according to the way the Earth's internal heat is dissipated. There are continental plates like North America made essentially of granite-related rocks, and oceanic plates like the Pacific made essentially of rocks with the composition of basalt.

11. A. Snir et al. 2015. 'The origin of cultivation and proto-weeds, long before Neolithic farming', *PLoS ONE*, **10**(7), e0131422. doi:10.1371/journal.pone.0131422.

12. L.M. Kiage, and K.-B. Liu. 2009. 'Palynological evidence of climate change and land degradation in the Lake Baringo area, Kenya, East Africa, since AD 1650', *Palaeogeography, Palaeoclimatology, Palaeoecology*, **279**, 60–72.

13. H. Seebens et al. 2017. 'No saturation in the accumulation of alien species worldwide', *Nature Communications*, **8**, article number 14435. https://doi. org/10.1038/ncomms14435/.

14. To add insult to injury, lionfish may have invaded as escapees from an aquarium. https://www.fisheries.noaa.gov/feature-story/impacts-invasive-lionfish/.

15. In fact, gold had been discovered several decades earlier in California, but the news of these discoveries had stayed local.

16. See Richard Peterson's article on Historynet at https://www.historynet.com/james-marshall-californias-gold-discoverer.htm (accessed November 2021).

17. A.N. Cohen. 2004. 'Invasions in the sea', *Park Science*, **22**, 37–41.

18. A.N. Cohen, and J.T. Carlton. 1998. 'Accelerating invasion rate in a highly invaded estuary', *Science*, **279**, 555–557.

19. A. Benjamin. 2019. 'The battle to save the world's biggest bumblebee from extinction'. *The Guardian*. https://www.theguardian.com/environment/2019/may/04/the-battle-to-save-the-worlds-biggest-bumblebee-from-european-invaders/.

20. M.A. Aizen et al. 2019. 'Coordinated global species-importation policies are needed to reduce serious invasions globally: The case of alien bumble bees in South America', *Journal of Applied Ecology*, **56**, 100–106.

21. T.-B. Yang et al. 2013. 'The Apple Snail *Pomacea canaliculata*, a novel vector of the rat lungworm *Angiostrongylus cantonensis*: Its introduction, spread and control in China', *Hawaii Journal of Medicine and Public Health*, **72**, 23–25. https://www.ncbi.nlm.nih.gov/pmc/articles/PMC3689487/.

22. A.B.R. Witt et al. 2017. 'A preliminary assessment of the extent and potential impacts of alien plant invasions in the Serengeti- Mara ecosystem, East Africa', *Koedoe*. **59**(1), a1426. https://doi.org/10.4102/koedoe.v59i1.1426/.

23. M. Samways. 1999. 'Translocating fauna to foreign lands, here comes the Homogenocene', *Journal of Insect Conservation*, **3**, 65–66.

24. Crutzen's improvised vision is brilliantly summarized in P.J. Crutzen. 2002. 'Geology of Mankind', *Nature*, **415**, 23. A longer, reflective account of this concept is in J.A. Thomas, M. Williams, and J. Zalasiewicz. 2020. *The Anthropocene: A Multidisciplinary Approach* (Cambridge: Polity Books).

25. M.J. Watson, and D.M. Watson. 2019. 'Post-Anthropocene conservation', *Trends in Ecology & Evolution*, **35**. doi:https://doi.org/10.1016/j.tree.2019.10.006.

CHAPTER 4

1. A. Mayor. 2011. *The First Fossil Hunters* (Princeton: Princeton University Press).

2. A. Gibbons. 2018. 'Strongest evidence of early humans butchering animals discovered in North Africa'. https://www.sciencemag.org/news/2018/11/strongest-evidence-early-humans-butchering-animals-discovered-north-africa (accessed December 2021).

3. J.A.J. Gowlett. 2016. 'The discovery of fire by humans: A long and convoluted process', *Philosophical Transactions of the Royal Society B* **371**, 20150164.

4. P. Brunton. 2009. 'Through Darwin's eyes. Australia played an important role in Charles Darwin's theory of evolution'. https://www.sl.nsw.gov.au/stories/through-darwins-eyes.

5. D. Cowley, and B. Hubber. 2000. 'Distinct creation: Early European images of Australian animals'. http://www3.slv.vic.gov.au/latrobejournal/issue/latrobe-66/t1-g-t2.html

6. T.H. Worthy. 2016. 'A case of mistaken identity for Australia's extinct big bird'. https://theconversation.com/a-case-of-mistaken-identity-for-australias-extinct-big-bird-52856.

7. D. Bustos et al. 2018. 'Footprints preserve terminal Pleistocene hunt? Human-sloth interactions in North America', *Science Advances*, **4**, eaar7621.

8. G.G. Politis et al. 2019. 'Campo Laborde: A Late Pleistocene giant ground sloth kill and butchering site in the Pampas', *Science Advances*, **5**, eaau4546.

9. Y. Mahli. et al. 2016. 'Megafauna function from the Pleistocene to the Anthropocene', *Proceedings of the National Academy of Sciences*, **113**, 838–846.

10. J.W. Laundré et al. 2010. 'The landscape of fear: Ecological implications of being afraid', *Open Ecology Journal*, **3**, 1–7.

11. See references in Y. Mahli. et al. 2016. 'Megafauna function from the Pleistocene to the Anthropocene', *Proceedings of the National Academy of Sciences*, **113**, 838–846.

12. Callum Roberts. 2007. *The Unnatural History of the Sea* (Wahington, DC: Island Press), and 2013. *Ocean of Life: The Fate of Man and the Sea* (New York: Penguin). Also, R.S. Steneck, and D. Pauly. 2019. 'Fishing through the Anthropocene', *Current Biology*, **29**, R942–R995.

13. W. Cornwall. 2019. 'Mountains hidden in the deep sea are biological hotspots. Will mining ruin them?' *Science*. doi:10.1126/science.aaz4600.

14. S. Guduff et al. 2018. 'A case study of the Walters Shoal in the South West Indian Ocean'. https://www.iddri.org/sites/default/files/PDF/Publications/Hors%20catalogue%20Iddri/201806-rapport%20walters%20shoalEN.pdf.

15. IUCN. 'Walters Shoal's marine fauna and flora—The Walters Shoal's summit: an algal reef!' https://science4highseas.wixsite.com/waltersshoal/single-post/2017/05/13/Walters-Shoals-marine-fauna-and-flora-The-Walters-Shoals-summit-an-algal-reef (accessed November 2021).

16. F. Marsac et al. 2019. 'Seamounts, plateaus and governance issues in the southwestern Indian Ocean, with emphasis on fisheries management and marine conservation, using the Walters Shoal as a case study for implementing a protection framework', *Deep Sea Research Part II*. https://doi.org/10.1016/j.dsr2.2019.104715.

17. Oldest cat ever. https://www.guinnessworldrecords.com/world-records/oldest-cat-ever (accessed November 2021)

18. We tell Ming's full story in *Skeletons: The Frame of Life* (Oxford University Press, 2018).

19. K.D. Bidle et al. 2007. 'Fossil genes and microbes in the oldest ice on Earth', *Proceedings of the National Academy of Sciences*, **104**, 13455–13460. https://www.pnas.org/content/104/33/13455/.

20. L. Trutnau, and R. Sommerlad. 2006. *Crocodilians: Their Natural History and Captive Husbandry* (Frankfurt am Main, Germany: Editiona Chimaira).

21. C.R. Poole, and B. Wade. 2019. 'Systematic taxonomy of the *Trilobus sacculifer* plexus and descendant *Globigerinoidesella fistulosa* (planktonic foraminifera)', *Journal of Systematic Palaeontology*, **17**, 1989–2030.

22. In fact, 468 more species of vertebrate have gone extinct in this interval, indicating the impacts of humans on life: G. Ceballos et al. 2015. 'Accelerated modern human–induced species losses: Entering the sixth mass extinction', *Science Advances*, **1**, e1400253. doi: 10.1126/sciadv.1400253.

23. M-X. Ling et al. 2019. 'An extremely brief end Ordovician mass extinction linked to abrupt onset of glaciation', *Solid Earth Sciences*, **4**, 190–198.

24. A good overview is given in P. Schulte et al. 2010. 'The Chicxulub asteroid impact and mass extinction at the Cretaceous-Paleogene boundary', *Science*, **327**, 1214–1218.

25. S.P.S. Gulick et al. 2019. 'The first day of the Cenozoic', *Proceedings of the National Academy of Sciences*, **116**, 19342–19351.

26. G. Ceballos et al. 2015. 'Accelerated modern human–induced species losses: Entering the sixth mass extinction', *Science Advances*, **1**, e1400253. doi: 10.1126/sciadv.1400253.

27. A.D. Barnosky et al. 2011. 'Has the Earth's sixth mass extinction already arrived?' *Nature*, **471**, 51–57.

28. M. Blaxter, and P. Sunnucks. 2011. 'Velvet worms', *Current Biology*, **27**, R238–240.

29. I. de S. Oliveira et al. 2012. 'A world checklist of Onychophora (velvet worms), with notes on nomenclature and status of names', *ZooKeys*, **211**, 1–70.

30. A. Sosa-Bartuano. et al. 2018. 'A proposed solution to the species problem in velvet worm conservation (Onychophora)', *UNED Research Journal*, 10, 193–197.

31. M.R. Smith, and J. Ortega-Hernández. 2014. '*Hallucigenia's* onychophoran-like claws and the case for Tactopoda', *Nature*, **514**, 363–366.

32. B. Morena-Brenes et al. 2019. 'The conservation status of Costa Rican velvet worms (Onychophora): Geographic pattern, risk assessment and

comparison with New Zealand velvet worms', *UNED Research Journal*, **11**, 272–282.

33. See also J. Thomas et al. 2020. *The Anthropocene: A Multidisciplinary Approach* (Cambridge: Polity Books).

34. J.P.G.M. Cromsigt, and M. Te Beest. 2014. 'Restoration of a megaherbivore: Landscape-level impacts of white rhinoceros in Kruger National Park, South Africa', *Journal of Ecology*, **102**, 566–575.

35. B. Farquhar. 2021. 'Wolf reintroduction changes ecosystem in Yellowstone'. https://www.yellowstonepark.com/things-to-do/wolf-reintroduction-changes-ecosystem (accessed November 2021).

36. P. Barkham. 2018. 'Dutch rewilding experiment sparks backlash as thousands of animals starve'. https://www.theguardian.com/environment/2018/apr/27/dutch-rewilding-experiment-backfires-as-thousands-of-animals-starve (accessed November 2021).

37. F. Sánchez-Bayo, and K.A.G. Wyckhuys. 2019. 'Worldwide decline of the entomofauna: A review of its drivers', *Biological Conservation*, **232**, 8–27.

38. Some estimates suggest as many as 5.5 million insects, and 1.5 million of these may be beetles, see N.E. Stork. 2018. 'How many species of insects and other terrestrial arthropods are there on Earth?' *Annual Review of Entomology*, **63**, 31–45.

CHAPTER 5

1. J.M.K.C. Donev et al. 2020. 'Energy education—Solar energy to the Earth'. https://energyeducation.ca/encyclopedia/Solar_energy_to_the_Earth (accessed 10 June 2020). The average radiation intensity that hits the edge of the Earth's atmosphere is 1367 watts per square metre. This is an average, remembering that half of the planet will be in darkness, and that the surface of the atmosphere is a sphere, and more energy will fall per square metre in the tropics than the poles. Hence the average of 340 watts per square metre.

2. X.-G. Zhu et al. 2008. 'What is the maximum efficiency with which photosynthesis can convert solar energy into biomass?' *Current Opinion in Biotechnology*, **19**, 153–159.

3. Y.M. Bar-On et al. 2018. 'The biomass distribution on Earth', *Proceedings of the National Academy of Sciences*, **115**, 6506–6511.

4. We describe this amazing diversity in *Skeletons: The Frame of Life* (Oxford University Press, 2018).

5. These are tiny worms, sometimes called roundworms, that are present in most of Earth's environments.

6. As we saw in Chapter 2, taxonomy defines a 'diagnosis' for all organisms, living and fossil, this being the key feature or features that distinguish each type of organism from another.

7. J.R. Burger, and T.S. Fritsoe. 2018. 'Hunter-gatherer populations inform modern ecology', *Proceedings of the National Academy of Sciences*, **115**, 1137–1139.

8. Air without its combustible component 'phlogiston', for which see Chapter 1. Of course, there is no such thing as phlogiston.

9. J.N. Galloway et al. 2013. 'A chronology of human understanding of the nitrogen cycle', *Philosophical Transactions of the Royal Society, Series B*, **368**, 20130120. http://dx.doi.org/10.1098/rstb.2013.0120.

10. Initially, the catalyst was the metal osmium, though as this is rare, when the process was scaled to industrial processes an iron-based catalyst was developed.

11. R. Carty. 2012. 'Casualty of war'. https://www.sciencehistory.org/distillations/magazine/casualty-of-war/.

12. F. Bretislav, and D. Hoffman. 2016. 'Clara Haber, nee Immerwahr (1870–1915): Life, work and legacy', *Zeitschrift fur anorganische und allgemeine Chemie*, **642**, 437–448. https://www.ncbi.nlm.nih.gov/pmc/articles/PMC4825402/.

13. Centro Internacional de Mejoramiento de Maíz y Trigo (CIMMYT).

14. Hannah Ritchie, and Max Roser. 2013 (most recent revision in 2021). 'Crop yields'. *OurWorldInData.org*. https://ourworldindata.org/crop-yields (accessed December 2021).

15. H. Ritchie, and M. Roser. 2017 (revised in 2018). 'Water use and stress', *OurWorldInData.org*. https://ourworldindata.org/water-use-stress (accessed December 2021).

16. Y.Y. Balega et al. 2012. 'Nikolai Semenovich Kardashev (on his 80th birthday)', *Physics-Uspekhi*, **55**, 63.

17. N.S. Kardashev. 1964. 'Transmission of information by extra-terrestrial civilizations', *Soviet Astronomy-AJ*, **8**, 217–221.

18. https://www.statista.com/statistics/265598/consumption-of-primary-energy-worldwide/ (accessed November 2021).

19. See Chapter 7, section 7.7.1 of GEA. 2012. *Global Energy Assessment—Toward a Sustainable Future* (Cambridge, UK and New York, NY, USA: Cambridge University Press, and Laxenburg, Austria: the International Institute for Applied Systems Analysis). http://www.iiasa.ac.at/web/home/research/

Flagship-Projects/Global-Energy-Assessment/Chapters_Home.en.html (accessed November 2021)

20. I. Dostrovsky. 1991. 'Chemical fuels from the sun', *Scientific American*, **265**, 102–107.

21. Energy use in metal production. https://publications.csiro.au/rpr/download?pid=csiro:EP12183&dsid=DS3 (accessed November 2021).

22. K.J. Ptasinski. 2016. *Efficiency of Biomass Energy: An Exergy Approach to Biofuels, Power and Biorefineries* (Hoboken, NJ: Wiley).

23. O. Heffernan. 2019. 'Seabed mining is coming—bringing mineral riches and fears of epic extinctions', *Nature*, **571**, 465–468.

24. F. Creutzig et al. 2015. 'Global typology of urban energy use and potentials for an urbanization mitigation wedge', *Proceedings of the National Academy of Sciences*, **112**, 6283–6288.

25. M. Dade-Robertson et al. 2017. 'Architects of nature: Growing buildings with bacterial biofilms', *Microbial Biotechnology*, **10**, 1157–1163. C.M. Heveran et al. 2020. 'Biomineralization and successive regeneration of engineered living building materials', *Matter*, **2**, 481–494.

26. 'Recyclable architecture—changing how we build, use, and re-use urban structures', *Urban Hub*. https://www.urban-hub.com/cities/recyclable-architecture-changing-how-we-build-use-and-re-use-urban-structures/ (accessed November 2021).

27. Aalto University. 2020. 'Building cities with wood would store half of cement industry's current carbon emissions', *ScienceDaily*, 2 November. www.sciencedaily.com/releases/2020/11/201102110010.htm (accessed November 2021).

28. M. Williams et al. 2022. 'Mutualistic Cities of the Near Future'. In J.A. Thomas, ed., *Altered Earth: Getting the Anthropocene Right* (Cambridge University Press).

29. R.I. McDonald et al. 2014. 'Water on an urban planet: Urbanization and the reach of urban water infrastructure', *Global Environmental Change*, **27**, 96–105.

30. Plant-e. 'Plant-E and IRNAS stand-alone plant-power device for IOT solutions'. https://www.plant-e.com/en/plant-e-and-irnas-stand-alone-plant-power-device-for-iot-solutions/ (accessed November 2021), and WEF. 'Q&A: How the roots of living plants can power your wifi'. https://www.weforum.org/agenda/2016/06/q-a-how-the-roots-of-living-plants-can-power-your-wifi/ (accessed November 2021).

31. J.-Y. Min et al. 2019. 'Leaf anatomy and 3-D structure mimic to solar cells with light trapping and 3-D arrayed submodule for enhanced electricity

production', *Scientific Reports*, **9**, article number 10273. https://www.nature.com/articles/s41598-019-46748-x. G. Sertic. 2015. 'Nature's blueprint: Solar cells inspired by plant cells', *Yale Scientific*. https://www.yalescientific.org/2015/11/natures-blueprint-solar-cells-inspired-by-plant-cells/ (accessed November 2021).

CHAPTER 6

1. R. Moss. 2017. 'The 1,500-year-old recipe that shows how Romans invented the burger'. *Museum Crush*. https://museumcrush.org/the-1500-year-old-recipe-that-shows-how-romans-invented-the-beef-burger/ (accessed November 2021).

2. Fairmount History. https://www.jamesdeanartifacts.com/fairmount-history.html/ (accessed November 2021).

3. McDonalds Statistics, Restaurant Count, Revenue Totals and Facts. 2021. https://expandedramblings.com/index.php/mcdonalds-statistics/ (accessed November 2021).

4. W. Steffen et al. 2007. 'The Anthropocene: Are humans now overwhelming the great forces of Nature?' *Ambio*, **36**, 614–621.

5. Statista. 'Number of Burger King restaurants worldwide from 2009 to 2020'. https://www.statista.com/statistics/222981/number-of-burger-king-restaurants-worldwide/ (accessed November 2021).

6. C. Andrews. 'July Fourth food! How many hot dogs and hamburgers are consumed in your state?' https://eu.usatoday.com/story/money/2019/06/17/july-4th-hot-dog-and-hamburger-consumption-by-state/39580323/ (accessed November 2021).

7. Statista. 'Average floor space of select quick service restaurants (QSRs) in the United States in 2016'. https://www.statista.com/statistics/587130/average-floor-space-qrs-us/ (accessed November 2021).

8. H. Ritchie. 2017. 'How much of the world's land would we need in order to feed the global population with the average diet of a given country?' https://ourworldindata.org/agricultural-land-by-global-diets (accessed December 2021).

9. Hyde and Rugg. 'Caesar on the aurochs'. https://hydeandrugg.wordpress.com/2014/06/17/caesar-on-the-aurochs/ (accessed November 2021).

10. C.J. Caesar, *De Bello Gallico*. Project Gutenberg; Everyman's Library version, 1915 edition, translated by W.A. MacDevitt.

11. R. Bollongino et al. 2012. 'Modern taurine cattle descended from small number of Near-Eastern founders', *Molecular Biology and Evolution*, **29**, 2101–2104.

12. V. Smil. 2011. 'Harvesting the biosphere: The human impact', *Population and Development Review*, **37**, 613–636.

13. R.A. Lawal et al. 2020. 'The wild species genome ancestry of domestic chickens', *BMC Biology*, **18**, article 13. doi:10.1186/s12915-020-0738-1.

14. H.L. Shrader. 1952. 'The Chicken of Tomorrow program: Its influence on "meat-type" production', *Poultry Science*, **31**, 3–10.

15. C.E. Bennett et al. 2018. 'The broiler chicken as a signal of a human re-configured biosphere', *Royal Society Open Science*, **5**, article number 180325. http://10.1098/rsps.180325.

16. B. Cole et al. 2015. 'Corine Land Cover 2012 for the UK, Jersey and Guernsey'. NERC Environmental Information Data Centre. https://doi.org/10.5285/32533dd6-7c1b-43e1-b892-e80d61a5ea1d.

17. G.D. Powney et al. 2019. 'Widespread losses of pollinating insects in Britain', *Nature Communications*, **10**, article number 1018. https://doi.org/10.1038/s41467-019-08974-9.

18. British Trust for Ornithology. 'Bird indicators'. https://www.bto.org/our-science/publications/developing-bird-indicators (accessed November 2021).

19. K. Gajewski et al. 2019. 'Human-vegetation interactions during the Holocene in North America', *Vegetation History and Archaeobotany*, **28**, 635–647.

20. C.M. Kennedy et al. 2019. 'Managing the middle: A shift in conservation priorities based on the global human modification gradient', *Global Change Biology*, **25**, 811–826. https://onlinelibrary.wiley.com/doi/pdf/10.1111/gcb.14549.

21. B.S. Halpern et al. 2015. 'Spatial and temporal changes in cumulative human impacts on the world's ocean', *Nature Communications*, **6**, article number 7615. https://doi.org/10.1038/ncomms8615.

22. L. Stephens et al. 2019. 'Archaeological assessment reveals Earth's early transformation through land use', *Science*, **365**, 897–902. http://globe.umbc.edu/archaeoglobe/.

23. PBS. 'Gabriel's conspiracy'. https://www.pbs.org/wgbh/aia/part3/3p1576.html (accessed November 2021).

24. Global Agriculture. 'Agriculture at a crossroads: Findings and recommendations for future farming'. https://www.globalagriculture.org/report-topics/meat-and-animal-feed.html (accessed November 2021).

25. Hannah Ritchie. 2017. 'How much of the world's land would we need in order to feed the global population with the average diet of a

given country?' https://ourworldindata.org/agricultural-land-by-global-diets (accessed December 2021).

26. Ritchie. 2017. 'How much of the world's land?' https://ourworldindata.org/agricultural-land-by-global-diets (accessed December 2021).

27. P. Alexander et al. 2016. 'Human appropriation of land for food: the role of diet', *Global Environmental Change*, **41**, 88–98.

28. P. Alexander et al. 2016. 'Human appropriation of land for food', *Global Environmental Change*, **41**, 88–98.

29. R.O. Amoroso et al. 2018. 'Bottom trawl fishing footprints on the world's continental shelves', *Proceedings of the National Academy of Sciences*, **115**, E10275–E10282.

30. N. Pacoureau et al. 2021. 'Half a century of global decline in oceanic sharks and rays', *Nature*, **589**, 567–571.

31. C.J.A. Bradshaw et al. 2008. 'Decline in whale shark size and abundance at Ningaloo Reef over the past decade: The world's largest fish is getting smaller', *Biological Conservation*, **141**, 1894–1905.

32. T.A. Branch. 2001. 'A review of orange roughy *Hoplostethus atlanticus* fisheries, estimation methods, biology and stock structure', *South African Journal of Marine Science*, **23**, 181–203. doi: 10.2989/025776101784529006.

33. K. Evans. 'The 230-year-old fish', *New Zealand Geographic*. https://www.nzgeo.com/stories/the-230-year-old-fish (accessed November 2021).

34. FAO. 1985. 'Freshwater aquaculture development in China'. Fisheries technical paper 215. Rome. ISBN 92-5-101,113-3.

35. T. Nakajima. 2019. 'Common carp aquaculture in Neolithic China dates back 8,000 years', *Nature Ecology and Evolution*, **3**, 1415–1418. https://www.nature.com/articles/s41559-019-0974-3.

36. J. Harland. 2019. 'The origins of aquaculture', *Nature Ecology and Evolution*, **3**, 1378–1379. https://doi.org/10.1038/s41559-019-0966-3

37. G. Sisma-Ventura et al. 2018. 'Tooth oxygen isotopes reveal Late Bronze Age origin of Mediterranean fish aquaculture and trade', *Scientific Reports*, **8**, article number 14086. doi:10.1038/s41598-018-32468-1.

38. M.S. Busana. 2018. 'Fishing, fish farming and fish processing during the Roman age in the Northern Adriatic: Literary sources and archaeological data', *Regional Studies in Marine Science*, **21**, 7–16.

39. R.C. Hoffmann. 2005. 'A brief history of aquatic resource use in medieval Europe', *Helgoland Marine Research*, **59**, 22–30.

40. J. Harland. 2019. 'The origins of aquaculture', *Nature Ecology and Evolution*, **3**, 1378–1379. https://doi.org/10.1038/s41559-019-0966-3. J.H. Primavera. 2006. 'Overcoming the impacts of aquaculture on the coastal zone', *Ocean & Coastal Management*, **49**, 531–545.

41. F. Teletchea, and P. Fontaine. 2014. 'Levels of domestication in fish: Implications for the sustainable future of aquaculture', *Fish and Fisheries*, **15**, 181–195.

42. F. Teletchea, and P. Fontaine. 2014. 'Levels of domestication in fish', *Fish and Fisheries*, **15**, 181–195.

43. A human definition of a fish that traditionally had no foodstuff value to humans.

44. J. Gould, and M. Ashour. 2017. 'A world of insecurity', *Nature*, **544**, S6–S7.

45. A. King. 2017. 'The future of agriculture', *Nature*, **544**, S21–S23.

46. K. Bourzac. 2017. 'Solar upgrade', *Nature*, **544**, S11–S13.

47. Bourzac. 2017. 'Solar upgrade', *Nature*, **544**, S11–S13.

48. O. Heffernan. 2017. 'A meaty issue', *Nature*, **544**, S18–20.

CHAPTER 7

1. G. Shelach. 2012. 'On the invention of pottery', *Science*, **336**, 1644.

2. C. Mora et al. 2011. 'How many species are there on Earth and in the ocean?' *PLoS Biology*, **9**, e100127.

3. C. Schulz. 2012. 'The story of László Bíró, the man who invented the ballpoint pen', *Smithsonian Magazine*. https://www.smithsonianmag.com/smart-news/the-story-of-laszlo-biro-the-man-who-invented-the-ballpoint-pen-30631082/ (accessed November 2021).

4. 'Fountain pen—History of fountain pens'. http://www.historyofpencils.com/writing-instruments-history/fountain-pen-history/ (accessed November 2021).

5. M. Popova. 'The surprising history of the pencil'. https://www.brainpickings.org/2013/06/24/history-of-the-pencil/ (accessed November 2021).

6. J. Zalasiewicz et al. 2017. 'Scale and diversity of the physical technosphere: A geological perspective', *The Anthropocene Review*, **4**, 9–22.

7. The Inorganic Crystal Structure Database: http://icsd.fiz-karsruhe.de/.

8. P.J. Heaney. 2017. 'Defining minerals in the age of humans', *American Mineralogist*, **102**, 925–926: commenting on R.M.Hazen, E.S. Grew, M.J. Origlieri, and R.T. Downs. 2017. 'On the mineralogy of the "Anthropocene Epoch"', *American Mineralogist*, **102**, 595–611.

9. P.K. Haff. 2014. 'Technology as a geological phenomenon: Implications for human well-being', *Geological Society of London, Special Publications* 395, 301–309. See also: P. Haff. 2019. 'The Technosphere and Its Physical Stratigraphic Record'. In J. Zalasiewicz, C.N. Waters, M. Williams, and C.P.

Summerhayes, eds., *The Anthropocene as a Geological Time Unit: A Guide to the Scientific Evidence and Current Debate* (Cambridge, UK: Cambridge University Press), 137–155.

10. E. Elhacham et al. 2020. 'Global human-made mass now exceeds all living biomass', *Nature*, **588**, 442–444.

11. Y.M. Bar-On et al. 2018. 'The biomass distribution on Earth', *Proceedings of the National Academy of Sciences*, **115**, 6506–6511.

12. J. Zalasiewicz et al. 2017. 'Scale and diversity of the physical technosphere: A geological perspective', *The Anthropocene Review*, **4**, 9–22.

13. Haff. 2019. 'The Technosphere and Its Physical Stratigraphic Record'. In J. Zalasiewicz, C.N. Waters, M. Williams, and C.P. Summerhayes, eds., *The Anthropocene as a Geological Time Unit* (Cambridge, UK: Cambridge University Press), 137–155.

14. S. Kriegman, D. Blackiston et al. 2020. 'A scalable pipeline for designing reconfigurable organisms', *Proceedings of the National Academy of Sciences*, **117**, 1853–1859.

15. In a commentary on the work by P. Ball. 2020. 'Living robots', *Nature Materials*. https://doi.org/10.1038/s41563-020-0627-6. The authors of the research simply call them 'biobots'.

16. J. Zalasiewicz et al. 2014. 'The technofossil record of humans', *Anthropocene Review*, **1**, 34–43.

17. J. Zalasiewicz. 2008. *The Earth After Us: The Legacy That Humans Will Leave in the Rocks* (Oxford, New York: Oxford University Press).

18. P.M. Kahara et al. 2019. 'Determination of selected heavy metals in tobacco tree shrubs growing around Dandora Dumpsite, Nairobi, Kenya', *Chemical Science International Journal*, **27**, 1–9. https://doi.org/10.9734/CSJI/2019/v27i330117.

19. E. Song'oro et al. 2019. 'Occurrence of highly resistant microorganisms in Ruai Wastewater Treatment Plant and Dandora Dumpsite in Nairobi County, Kenya', *Advances in Microbiology*, **9**, 479–494. https://doi.org/10.4236/aim.2019.95029.

CHAPTER 8

1. A. Estrada et al. 2017. 'Impending extinction of the world's primates: Why primates matter', *Science Advances*, **3**, e1600946. https://advances.sciencemag.org/content/3/1/e1600946.

2. J. Turner. 2021. 'Palm oil free list'. https://www.ethicalconsumer.org/palm-oil/palm-oil-free-list (accessed November 2021).

3. N. Coca. 2018. 'Despite government pledges, ravaging of Indonesia's forests continues'. https://e360.yale.edu/features/despite-government-pledges-ravaging-of-indonesias-forests-continues (accessed November 2021).

4. WWF. 'Southeastern Asia: Indonesia and Malaysia'. https://www.worldwildlife.org/ecoregions/im0102 (accessed November 2021).

5. J.H. Schwartz et al. 1995. 'A review of the Pleistocene hominoid fauna of the Socialist Republic of Vietnam (excluding Hylobatidae)', *Anthropological Papers of the American Museum of Natural History*, **76**, 1–24.

6. A. Reese. 2017. 'Newly discovered orangutan species is also the most endangered', *Nature*, **551**, 151. doi:10.1038/nature.2017.22934.

7. S. Leahy. 2019. 'Hydroelectric dam threatens to wipe out world's rarest ape'. https://www.nationalgeographic.com/animals/2018/08/tapanuli-orangutan-rarest-ape-threatened-dam-news/ (accessed November 2021).

8. D. Gaveau. 2019. 'Satellites check oil palm expansion in Borneo'. Scidev.net. https://www.scidev.net/asia-pacific/opinions/satellites-check-oil-palm-expansion-in-borneo/ (accessed November 2021).

9. In extreme circumstances, 'Eek!' could be brought into play.

10. See the magnificent essay 'The Orangutans are Dying' in his collection *A Slip of the Keyboard*.

11. The 'songs of Chu' is from the kingdom of that name in southern China that existed between the eighth to third centuries BCE.

12. Quoted on p. 218 of P. Zhang. 2015. 'Good gibbons and evil macaques: A historical review on cognitive features of non-human primates in Chinese traditional culture', *Orimates*, **56**, 215–225.

13. Robert van Gulik published an entire monograph in 1967, *The Gibbon in China: An Essay in Chinese Animal Lore*.

14. P. Zhang. 2015. 'Good gibbons and evil macaques', *Orimates*, **56**, 215–225.

15. P. Fan. 2017. 'The past, present and future of gibbons in China', *Biological Conservation*, **210**, 29–39.

16. V.V. Venkataraman. 2015. 'Solitary Ethiopian wolves increase predation success on rodents when among grazing gelada monkey herds', *Journal of Mammalogy*, **96**, 129–137.

17. Atlas Obscura. 'Spider monkeys of Tikal'. https://www.atlasobscura.com/places/spider-monkeys-of-tikal (accessed November 2021).

18. A. Estrada. et al. 2012. 'Agroecosystems and primate conservation in the tropics: A review', *American Journal of Primatology*, **74**, 696–711.

19. L.X. Lokschin et al. 2007. 'Power lines and Howler monkey conservation in Porto Allegre, Rio Grande do Sul, Brazil', *Neotropical Primates*, **14**, 76–80.

20. F.Z. Teixeira et al. 2013. 'Canopy bridges as road overpasses for wildlife in urban fragmented landscapes', *Biota Neotropica*, **13**. https://doi.org/10.1590/S1676-06032013000100013.

21. X.-P. Song et al. 2018. 'Global land change from 1982 to 2016', *Nature*, **560**, 639–643.

22. P.G. Curtis et al. 2015. 'Classifying drivers of global forest loss', *Science*, **361**, 1108–1111.

23. Y.M. Bar-On et al. 2018. 'The biomass distribution on Earth', *Proceedings of the National Academy of Sciences*, **115**, 6506–6511.

24. E.O. Wilson. 2016. 'Half Earth: Our planet's fight for life'. https://www.half-earthproject.org/.

25. E. Dinerstein et al. 2017. 'An ecoregion-based approach to protecting half the terrestrial realm', *BioScience*, **67**, 534–545.

26. B. Napoletano. 2018. 'Half-Earth: A biodiversity "solution" that solves nothing'. https://climateandcapitalism.com/2018/10/02/half-earth-a-biodiversity-solution-that-solves-nothing/ (accessed November 2021).

27. Wildlands network. https://wildlandsnetwork.org/wildways/eastern/ (accessed November 2021).

28. B. MacKaye. 1921. 'An Appalachian Trail: A project in regional planning', *Journal of the American Institute of Architects*, **9**, 325–330.

29. G.D. Nelson. 2019. 'An Appalachian Trail: A Project in Regional Planning'. https://placesjournal.org/article/an-appalachian-trail-a-project-in-regional-planning/?cn-reloaded=1 (accessed November 2021).

30. For example, one in eight British households has no garden. https://www.ons.gov.uk/economy/environmentalaccounts/articles/oneineightbritishhouseholdshasnogarden/2020-05-14 (accessed November 2021).

31. We describe such cities in our article: M. Williams et al. 2022. 'Mutualistic Cities of the Near Future'. In J.A. Thomas, ed., *Altered Earth: Getting the Anthropocene Right* (Cambridge University Press).

32. J. Owen. 2010. *Wildlife of a Garden: A Thirty-Year Study* (Royal Horticultural Society).

33. Buglife. 'Help Buglife save the planet'. https://www.buglife.org.uk/ (accessed November 2021).

34. Trees for Life. https://treesforlife.org.uk/dundreggan/ (accessed November 2021).

35. Carrifran wildwood. https://carrifran.bordersforesttrust.org/about/what-we-have-achieved/ecological-restoration/ (accessed November 2021).

36. A. Squires. Updated 2017. *Cloud Wood. A History and Natural History of Ancient Leicestershire Woodland* (Leicestershire and Rutland Wildlife Trust). https://www.lrwt.org.uk/sites/default/files/2020-02/cloud%20wood%20booklet%20for%20website%20feb2018_0.pdf (accessed November 2021)

37. T. Ireland. 2019. 'The artificial meat factory—the science of your synthetic supper'. https://www.sciencefocus.com/future-technology/the-artificial-meat-factory-the-science-of-your-synthetic-supper/ (accessed November 2021).

38. A.M. Cisneros-Montemayor et al. 2016. 'A global estimate of seafood consumption by coastal Indigenous peoples', *PLoS ONE*, **11**, e0166681. https://doi.org/10.1371/journal.pone.0166681.

39. C.R. Menzies, and C.F. Butler. 2007. 'Returning to selective fishing through Indigenous fisheries knowledge: The Example of K'moda, Gitaała Territory', *American Indian Quarterly*, **31**, 441–464.

40. IUCN. 'Mangrove restoration'. https://www.iucn.org/theme/forests/our-work/forest-landscape-restoration/mangrove-restoration (accessed November 2021).

41. K. Ervita, and M.A. Mrfai. 2017. 'Shoreline change analysis in Demak, Indonesia', *Journal of Environmental Protection*, **8**, 940–955.

42. EcoShape. 'Building with nature'. https://www.ecoshape.org/en/projects/building-with-nature-indonesia/ (accessed November 2021).

43. J.H. Primavera. 2006. 'Overcoming the impacts of aquaculture on the coastal zone', *Ocean & Coastal Management*, **49**, 531–545.

44. WWF. 'Mai Po wetland habitat'. http://awsassets.wwfhk.panda.org/downloads/gei_wai.pdf (accessed November 2021).

45. P.L. Reynolds. 2018. 'Seagrass and seagrass beds'. https://ocean.si.edu/ocean-life/plants-algae/seagrass-and-seagrass-beds (accessed November 2021).

46. M. Waycott et al. 2009. 'Accelerating loss of seagrasses across the globe threatens coastal ecosystems', *Proceedings of the National Academy of Sciences*, **105**, 12377–12381.

47. Seagrass ocean rescue. https://www.projectseagrass.org/seagrass-ocean-rescue/ (accessed November 2021).

48. Chesapeake Bay Foundation. 'Plants of the Chesapeake'. www.cbf.org/about-the-bay/more-than-just-the-bay/Chesapeake-plants/bay-grasses.html (accessed November 2021).

49. About Papahānaumokuākea. https://www.papahanaumokuakea.gov/new-about/ (accessed November 2021).

50. Seabirdwatch. https://www.zooniverse.org/projects/penguintom79/seabirdwatch/about/research (accessed November 2021).

51. Fishing in the past. https://www.zooniverse.org/projects/anneoverduin/fishing-in-the-past/about/research (accessed November 2021).

52. P.B. Moyle, and M.A. Moyle. 1991. 'Introduction to fish imagery in art', *Environmental Biology of Fishes*, **31**, 5–23.

53. F. Sommer, and F. Backhed. 2013. 'The gut microbiota—masters of host development and physiology', *Nature Reviews Microbiology*, **11**, 227–238.

54. R. Sender et al. 2016. 'Revised estimates for the number of human and bacterial cells in the body', *PLoS Biology*, **14**, e1002533. https://www.ncbi.nlm.nih.gov/pmc/articles/PMC4991899/.

55. E.D. Sonnenburg, and J.L. Sonnenburg. 2019. 'The ancestral and industrialized gut microbiota and implications for human health', *Nature Reviews Microbiology*, **17**, 383–390.

56. A. Tomova. et al. 2019. 'The effects of vegetarian and vegan diets on gut microbiota', *Frontiers in Nutrition*, **6**, 47. doi: 10.3389/fnut.2019.00047.

57. The modern Leviathan is eloquently described by Michel Callon and Bruno Latour in 'Unscrewing the big Leviathan: How actors macro-structure reality and how sociologists help them do so.' In K. Knorr-Cetina and A.V. Cicoural, eds., *Advances in Social Theory and Methodology: Toward an Integration of Micro- and Macro-Sociologies* (Routledge and Keegan Paul, 1981), 277–303.

INDEX